L.E.M.O.N.S. - BIASES IN CREATION

BY THOMAS. W. GUTHRIE

L.E.M.O.N.S. – Biases in Creation

ISBN: 979-8-9988546-0-6

Published by Thomas W. Guthrie
Land O' Lakes, Florida 34638

Dedication:

This book is dedicated to my family: my wife, Sandy; my children, Tara, Taylor, Tori, and their partners; my grandchildren, Callie, Seth, Taynori. Each of these family members have been important in teaching me to clearly understand the necessity of conveying the Gospel of Jesus Christ to all who may hear. Authorship of my books is one way of conveying the Gospel of Jesus Christ.

TABLE OF CONTENTS

PREFACE

L.E.M.O.N.S. – Biases in Creation at its core is focused upon the Bible and what is presented therein regarding creation of the universe. "L" stands for the first item referenced in the book of Genesis regarding creation – LIGHT. "E" represents the creation of the – EARTH. "M" is for – MAN. "O" is for outer-space. "N" addresses the recreation of the universe – NEW CREATION. "S" represents the subtleties and awe-inspiring aspects of creation, the – SUBLIME. The letters of the acronym LEMONS are addressed each in a separate chapter. Generally each chapter of the acronym also include representation of a scientific bias. And, persuasion for acceptance of a biblical bias is attempted.

The intent of this book is to present the biases as to the creation of the universe from a scientific bias, a biblical bias, a theological bias (not necessarily the same as a biblical bias) and the author's bias. The author's bias is underpinned in the biblical bias. Accordingly, such is the overtone of the book. There is no "proof" any of the biases presented are correct. There are points where biases overlap. And, while there is no "proof" as to the correctness of any of the biases; it is the author's desire that the biases presented herein allow the reader to contemplate the ramifications of each and steer towards appropriating the biblical bias as the basis for their worldview.

"The heavens declare the glory of God; and the firmament sheweth his handywork" - Psalm 19:1

CHAPTER 1

BASELINE

When considering creation we view the majesty of it, whether: the trees, the grass, a soft brown rabbit in the field, the bold red cardinal, a majestic bull elk bugling on the edge of the forest, the bright blue sky, clouds, mountains, deserts, seas, rain, snow, the sun, the stars, the development of a baby in the womb, children as they grow, man as he ages, a multitude of other things that draw our attention and even death. In fact, everything we can hear, see, taste, smell, touch or think is related to creation. In all this there is no denying the existence of creation but only the who or what, and the when.

There have been may books written, many debates about the who or what and even many books and debates about the when. The author believes what the Bible says about Jesus of Nazareth being the Christ and as a Christian so professes openly. Accordingly there is no debate herein about the "who", of creation as the Bible states; *"In the beginning God created..."* [1]. Therefore, the focus will identify the "when". And, in doing so some loose ends that many Christians ponder will be tied up. However, before moving to these issues a basis must be established.

Before the issues are presented in depth and detail there is helpful that you believe the Bible to be scripture. And, as scripture it is intended for our benefit as 2 Timothy 3:16, 17

---

1   Genesis 1:1, King James Version. Note that all scripture (Bible) references throughout are from the Kings James Version, unless otherwise noted.

states, *"All scripture is given by inspiration of God, and is profitable for doctrine, for reproof, for correction, for instruction in righteousness: (v. 16) That the man of God may be perfect, throughly furnished unto all good works. (v.17)"* In addition to knowing that scripture is for our benefit you must also understand that all proper understanding of scripture begins by recognizing and accepting the 'plain text' – that is the words that are written on the page as they are. Most certainly there are many deeper understandings than just the words themselves. But, if you cannot and do not accept and understand the plain text; the deeper understandings can be (and usually are) perverted.

As with any story, the story of the Bible is conveyed to the reader by the words on the page – the plain text. How can you read, little less understand the story if you do not accept the words on the page for what it says? In God's shown love for the world he has written a story for us to read, understand and accept. Why would you read such a wonderful story and throw it away for doctrines of men?

When reading and discussing scripture it is most important to focus upon the plain text (the words written) and the context of the plain text. While there are undoubtedly hidden nuggets of deeper wisdom in scripture; without focusing upon the plain text and the context of the plain text the main point of scripture is lost. God has taken time to reveal himself to his creation through scripture. Accordingly, it is illogical and irrational to discount the plain text.

In fact the plain text is the basis for any analysis and interpretation of scripture. Understanding the plain text is paramount. And though there are elements of poetic language, cultural euphemisms and allegory these elements are readily discernible in context. Whether reading about the exploits of the patriarchs or the kings in points of history or points of religion it is important to know what the words say and what the word says.

A key example of the plain text is one the author uses frequently when discussing the issue; John 11:35 "Jesus wept." Although there is the openness of interpretation as to why Jesus wept; the words say *"Jesus wept"*...he cried...or as Dictionary.com defines wept: *"simple past tense and participle of weep"...*"weep: to express grief, sorrow, or any overpowering emotion by shedding tears; shed tears; cry..."*

The entire basis of Christianity, Christian belief, Christian theology and one's Christian daily walk is the plain text of the Bible:

> *"In the beginning God created the heaven and the earth."*(Genesis 1:1);...
>
> *"And the children of Israel went into the midst of the sea upon the dry ground; and the waters were a wall unto them on their right hand, and on their left."*(Exodus 14:22);...
>
> *"Now the birth of Jesus Christ was on the wise; When as his mother Mary was espoused to Joseph, before they came together she was found with child of the Holy Ghost."*(Matthew 1:18);...
>
> *"Jesus, when he had cried again with a loud voice, yielded up the ghost."*(Matthew 27:50);...
>
> *"The Jews therefore, because it was the preparation, that*

11

the bodies should not remain upon the cross on the sabbath day. (for that sabbath day was an high day.) besought Pilate that their legs might be broken, and that they might be taken away. (v.31) Then came the soldiers, and brake the legs of the first, and of the other which was crucified with him. (v. 32) But when they came to Jesus, and saw that he was dead already, the brake not his legs: (v. 33) But one of the soldiers with a spear pierced his side, and forthwith came there out blood and water. (v. 34) And he that saw it bare record and the record is true... (v. 35)" (John 19: 31-35);...

"Now in the place where he was crucified there was a garden; and in the garden a new sepulchre, wherein was never man yet laid. (v. 41) There laid they Jesus...(v. 42)" (John 19: 41-42);...

"And when the sabbath was past, Mary Magdalene, and Mary the mother of James, and Salome, had bought sweet spices, that they might come and anoint him. (v. 1) And very early in the morning the first day of the week, they came unto the sepulchre at the rising of the sun. (v. 2) And they said among themselves, Who shall roll us away the stone from the door of the sepulchre? (v. 3) And when they looked, they saw that the stone was rolled away: for it was very great. (v. 4) And entering into the sepulchre, they saw a young man sitting on the right side, clothed in a long white garment; and they were affrighted. (v. 5) And he saith unto them, Be not affrighted: ye seek Jesus of Nazareth, which was crucified: he is risen; he is not here: behold the place where they laid him. (v. 6)" (Mark 16: 1-6);...

"The former treatise have I made, O Theophilus, of all

that Jesus began both to do and teach, (v. 1) Until the day in which he was taken up, after that he through the Holy Ghost had given commandments unto the apostles whom he had chosen: (v. 2) To whom also he shewed himself alive after his passion by many infallible proofs, being seen of them forty days, and speaking of the things pertaining to the kingdom of God: (v. 3) And, being assembled together with them, commanded them that they should not depart from Jerusalem, but wait for the promise of the Father, which, saith he, ye have heard of me. (v. 4) For John truly baptized with water; but ye shall be baptized with the Holy Ghost not many days hence. (v. 5) When they therefore were come together, they ask of him, saying, Lord, wilt thou at this time restore again the kingdom of Israel? (v. 6) And he said unto them, It is not for you to know the times or the seasons, which the Father hath put in his own power. (v. 7) But ye shall receive power, after that the Holy Ghost is come upon you: and ye shall be witnesses unto me both in Jerusalem, and in all Judaea, and in Samaria, and unto the uttermost part of the earth. (v. 8) And when he had spoken these things, while they beheld, he was taken up: and a cloud received him out of their sight. (v. 9) And while they looked steadfastly toward heaven as he went up, behold, two men stood by them in white apparel; (v. 10) Which also said, Ye men of Galilee, why stand ye gazing up into heaven? this same Jesus, which is taken up from you into heaven, shall so come in like manner as ye have seen him go into heaven. (v. 11)" (Acts 1: 1-11)

Creation, the deliverance of the Jewish people from Egyptian bondage, and the Gospel of Jesus – his birth, death, burial, resurrection and future return were all quoted above in plain text and designed to be read and understood in plain text.

13

So, as we move forward: focus on the plain text of the Bible. Review the plain text presented herein and validate it with the plain text of the Bible.

CHAPTER 2
THEOLOGICAL NON-UNIFORMITY

Even upon not contesting creation as being a act of God, there are multiple positions about when creation took place and how far along we are on the time frame from the origination. The four predominate positions are the old earth theory, the gap theory, the 'day-age' theory (both the gap theory and the 'day-age' theory version of the old earth theory and the young earth position. The old earth theory postulates that we are billions of years along the timeline since creation. The gap theory also postulates that we are billions of years along the timeline since creation; but, the billions of years is inserted in "a gap" between verses 1 and the first day of creation as related in verse 3 of the biblical reference of Genesis 1. The day-age theory postulates that each individual day of creation is not a twenty-four hour day but an 'age' of indeterminate length. The young earth position uses the biblical references of creation literally.

It should be noted that each of the four positions is (or was) supported by a variety of notable theologians. Divergence of positions, even that of allegorical interpretation, existed even among those of Christianity deemed "early church fathers". For example:

> Justin Martyr (100 – 165 AD) had a view of creation that it was over a long time frame using 2 Peter 3:8 as basis, *"...that one day is with the Lord as a thousand years, and a thousand years as one day."* [2]

2http://creationwiki.org/Justin_Martyr

Origen of Alexandria (184 – 253 AD) viewed the 6 days of creation as allegorical but, the instant creation as literal[3]

Eusebius (260/265- 339/340 AD) viewed the six days of creation literally and complied a history establishing creation as 3,184 years at the time of Abraham's birth [4]

Other authors studying the early church fathers and their approach to creation have observed that the early church fathers focused their attention not to the issue of the timing of creation but, perceived weightier issues associated with creation such as the concepts of 'ex nihilo' and the Trinity. Ex nihilo is Latin for 'out of nothing' wherein the idea is that God created the heavens and earth from elemental particles which were not preexistent but, were created at the time of recorded creation...in the beginning. The Trinity is the understanding that God is a single Godhead represented by God, the Father, God, the Son and God, the Holy Spirit. These issues were mainly explored by the early church fathers within their positioning and beliefs related to Christology (the study of Christ) and Eschatology (the study of the end times). Although many authors – whether supporters of young earth or supporters of old earth - attempt to use the early church fathers as a springboard to support a position on creation ; ultimately, there is no unanimity or even consensus among the early church fathers regarding the timing of creation.

Theology in its broadest sense does not exist in a vacuum within society. Likewise, in a narrow subject matter such as creation; theology does not exist in a vacuum within society. The

---

3https://creation.com/origen-origins-and-allegory
4 https://www.historyofinformation.com/detail.php?entryid=3804

subject of creation is and has been evaluated by all religions, most philosophies and society in general. Why are we here? How did we get here? Questions residing within the soul of mankind.

Not being within a vacuum, Jewish theologians debated their positions on creation against other viewpoints. Viewpoints such as those of the Greek philosophers Plato and Aristotle (Platonism) as well as contrary viewpoints within Judaism. With the advent of Christianity, Christian theologians debated and currently debate against prior philosophies, current philosophies, other theologies and scientific thought. Debate within this context as being not arguments; but persuasive discourse presented attempting to alter a contrary viewpoint to accept one's own viewpoint.

Surprisingly, during the period of the Christian Reformation in the 16th century and the prior diverse viewpoints on creation gelled into a broad consensus of a literal six days of creation. Surprisingly because the change of viewpoint occurred roughly 100 years after Copernicus established the basis of planetary orbital mechanics. It was postulated and observed that the earth and planets moved around the sun. At the time of the Reformation, the earth was no longer the center of the solar system.

Although belief in long time frames regarding creation have existed in history, it was not until the advent of paleontology within the scientific realm that the old earth theory became dominate in theology. At this point in history, old earth creation is the most widely held theory.

Thomas W. Guthrie

CHAPTER 3
SCIENTIFIC OBSERVATIONS

Science has developed four key avenues of observation for, what is believed in the scientific community, understanding creation time frames: biology (study of life), paleontology (study of the fossil record), geology (study of the physical structure of the earth) and cosmology (study of the origin and development of the universe). Each of these avenues, scientific disciplines, are dominated by a long time frame point of view.

Biology -

Man has throughout history pondered life and its creation. Not only the philosophical ascertaining of life - why but, how. Observation of human birth and death, the planting and growing of plants and animal husbandry all provided insights. However, through progression in history from the Age of Enlightenment (c. 1650 ad.) onward the coalescence of biological science as a discipline of strictly secular nature emerged.

In the mid 1800's biological science took on a new course. And with the published work of Charles Darwin, The Origin of Species in 1859 the new course was paved within the minds of both the scientific community and the people at large. The postulation of Darwin's observation and study was that individual species adapted physically to changing needs in their environment in order to perpetuate survival. This adaptation was referred to as a species 'evolving'. Thus, the theory of evolution was introduced into the biological sciences. Although Darwin did not postulate a broad species to species evolution in The

19

Origin of Species, the concept (and acceptance) of species to species evolution did soon thereafter become an anchor of biological science.

Evolution is in theory relegated as to taking place over long periods of time. Intraspecies evolution in many cases is perceived to take hundreds to thousand of years. Species to species evolution is perceived to take millions to billions of years. In any case evolution is a long-term event.

Geology -

Mankind has been observing his surroundings since the beginning of his existence. The earth – the ground - being of particular interest. Mountains, valleys, cliffs, hill, waterfalls, rivers, lakes and seas all provide points of observation with potentially different speculation as to the 'why' and 'how' of the items being observed. As man began to use the resources of the earth for survival and advancement, the complexity of the geology became more intriguing. Especially as resources were mined from underground.

Within geology there are three basic 'rock' types: sedimentary, igneous and metamorphic. Sedimentary rocks are those derived from the sediment of oceans, lakes or rivers. Igneous rocks are those associated with volcanic or magma events. Metamorphic rocks are rocks -either sedimentary or igneous – that are changed or metamorphosed by heat and pressure.

In times past even as now someone observing mountains and valleys will notice the layering or strata of different 'earth'. These layers may be massively thick or relatively thin. They may also be a combination of different types of strata;

sedimentary, igneous or metamorphic. From a common sense prospective the younger strata should be on top of or nearer to the surface of the earth than older strata. Yet,in some instances metamorphic rock may be situated in a strata closer to the surface than sedimentary strata. Regardless, the current predominate geological theory – evolutionary geology – theorizes geological events especially on a global scale take a long time to 'evolve' and deems the earth to be billions of years old.

Paleontology -

The study of the "fossil record" contained within the various geological strata has added to the theory of an earth with a very long history to date. As with other observations of the earth, man has observed fossils. It is noted that the Greeks as early as Xenophanes (570 – 480 b.c.) recognized and wrote about marine fossils [5] By the late 1700's paleontology began to emerge within the scientific community as a separate discipline. Although a portion of a dinosaur was found in England in 1676 [6] it wasn't properly identified for what it was until the early 1800's.

The discovery and cataloging of the various species of dinosaurs added to the theory of an aged earth. This is due to finding different species within different geological strata. There are three basic geological times associated with dinosaurs: Triassic (237 – 201 million years ago) , Jurassic (201 – 145 million years ago) and Cretaceous (145 – 66 million years ago). [7] As can be seen by the dates associated with the geological time frames of the dinosaurs, the earth is viewed as being in excess of

---

5   https://en.wikipedia.org/wiki/History_of_paleontology
6   https://www.dinosaurreport.com/first-dinosaur-discovered/
7   https://www.thoughtco.com/the-three-ages-of-dinosaurs-1091932

hundreds of millions of years old.

Cosmology –

The awe and wonderment expressed by man over the ages when contemplating the stars is staggering. Even with the naked eye; the vastness of the heavens, the innumerable points of light and the beauty draws our attention to creation. Even though man observed the sun, moon and 'wanderers' (the planets) throughout man's early history it wasn't until the time of Nicolaus Copernicus in the early-1500's that man's focus on the heavens changed. Copernicus theorized and that the sun was the center of the universe and not the earth. However, it was later observed that this theory was only partially correct in that the sun was the center of the solar system and not the universe. With the invention of the telescope in 1608 by Hans Lippershay (or Lipperhay) [8] the field of astronomy evolved. Theory and observation by Johannes Kepler (1571 – 1630), Galileo Galilei (1564 – 1642) and Isaac Newton (1642 - 1726) continued to refine solar system orbital mechanics and further man's understanding and inquiry into the universe. Orbital mechanics was sufficient for understanding our solar system but ineffective for inquiry into the broader universe.

Modern cosmology can be traced back to English theologian Robert Grosseteste and his thesis De Luce (on light) published in 1225. [9] Grosseteste's thesis, *"described the birth of the universe in an explosion and the crystallization of matter to*

8   https://www.space.com/21950-who-invented-the-telescope.html#:~:text=Hans%20Lippershey%2C%20credited%20with%20invention%20of%20the%20telescope.,a%20device%20that%20could%20magnify%20objects%20three%20times.
9   https://en.wikipedia.org/wiki/History_of_the_Big_Bang_theory#:~:text=The%20history%20of%20the%20Big%20Bang%20theory%20began,and%20refinements%20to%20the%20basic%20Big%20Bang%20model.

form stars and planets in a set of nested spheres around Earth. "
[10] So, even though the scientific advancement of the telescope enhanced study of the universe; an original basis for cosmology took root at an earlier time from within theology. Theological inquiry with a scientific purpose moved from the nested spheres concept to an expanding universe in 1927 by Belgian theologian, Georges Lemaitre. [11] However, Lemaitre's thesis retained Grosseteste's basis of an explosion as the creative event for the cosmos. It is from Lemaitre s thesis that the currently accepted "Big Bang" theory emanates.

Edwin Hubble, in 1927, theorized along similar lines as Lemaitre and subsequent observations based on his and Lemaitre's theories purport to validate an expanding universe based on a concept referred to as 'red shift'. It is from Lemaitre and Hubble that current cosmological dating is based. In the year 2020, the age of the universe was estimated to be 13.77 billion years old, give or take 40 million years [12]

Summary -

Observation by the four defined key scientific disciplines – biology, geology, paleontology and cosmology stand in basic agreement that creation has been a long term event; standing now at billions of years since inception. But, are the observations – or more accurately – the interpretation of the observations accurate? After all, the assumptions of observations are based upon theory. And, within scientific protocols; theory without duplication of experimentation and consistent results remains – theory.

---

10 Ibid.

11 Ibid

12 https:/www.livescience.com/universe-expansion-atacama-hubble-constant-measurement.html

Thomas W. Guthrie

CHAPTER 4
RETURN TO BASELINE

In Chapter 1 herein the author directed your attention to the baseline that the Bible and contents therein are valid. However, is the validity of the Bible ascertainable beyond the opinion and belief of the author?

Two key elements of scripture that support its validity are history and prophecy. According to the Christian Apologetics and Research Ministry (CARM); "The Bible was written over approximately 1600 years in three different languages, on three different continents, by 40 different authors." [13] This initial point of data sets the basis for evaluating the validity of the Bible. It is because of the multiplicity of authors and the lengthy time over which it was written that the allows a cross-reference and validation of events contained therein. This is true of both the history aspect and prophecy aspect of scripture.

Scripture, as compiled in the Bible is a compendium of history as related to the geographic area known as the Middle East or at an earlier time, Asia Minor. Some such figures of history referenced outside of scripture which are also referenced within scripture are: Nebuchadnezzar, king of Babylon; the Chaldean, successive empire to Babylon; Ahasuerus, king of Persia; Artaxerxes, king of Persia; Darius, king of Persia; Cyrus, king of Persia; Caesar Augustus, Emperor of Rome; Herod (the Great), Roman-appointed king of Judea and Jesus of Nazareth, the Messiah.

---

13 https://carm.org/the-bible/when-was-the-bible-written-and-who-wrote-it/

Thomas W. Guthrie

Although in scripture these historical figures are referenced from the viewpoint of the Hebrew (Jewish) people, the fact of their interactions is verified by history. For example, the Jewish captivity and exile in 586 BC (BCE), also known as the Babylonian diaspora, to Babylon after war between the Jewish kingdom of Judah and Nebuchadnezzar, king of Babylon has been found to be referenced through archaeological finds. The initial war in 597 BC (BCE) is referenced in the Babylonian Chronicles, a series of stone tablets recording major events in Babylonian history. [14]

Many of the Psalms in the book of Psalms of the Old Testament is generally attributed to King David. David (not yet king) is most recognized as the youth with a sling who slew the giant, Goliath described in the book of 1 Samuel, Chapter 17; and as king, committing adultery with Bathsheba as described in the book of 2 Samuel, chapter 12. David is mentioned in scriptures in the book of Ruth, the book of 1 Samuel, the book of 2 Samuel, the book of 1 Kings, the book of 2 Kings, the book of 1 Chronicles, the book of 2 Chronicles, the book of Ezra, the book of Nehemiah, the book of Psalms, the book of Proverbs, the book of Ecclesiastes, the book of Song of Songs, the book of Isaiah, the book of Jeremiah, the book of Ezekiel, the book of Hosea, the book of Amos, the book of Zechariah, the book of Matthew, the book of Mark, the book of Luke, the book of John, the book of Acts, the book of Romans, the book of 2 Titus, the book of Hebrews, the book of Revelation.

David is mentioned or referenced in 28 books of scripture; yet, from a secular viewpoint, throughout most of modern history, King David was viewed as a mythological figure and not a historical figure. The Bible though clearly declares King David as a historical figure. However, it wasn't until the discovery of

14 https://en.wikipedia.org/wiki/Babylonian_Chronicles

the Tel Dan inscription in 1993 ascribed to a 9th century BC (BCE) inscription in stone that directly references "the House of David"[15] that secularists assented to King David as a historical figure.

An especially intriguing interrelationship between history and prophecy is referenced in the Old Testament book of the Bible, Daniel. Daniel was one of the Jewish youth of wisdom that was exiled to Babylon under the command of Nebuchadnezzar, king of Babylon. According to the book of Daniel, Chapter 2, verses 1-49 Nebuchadnezzar had a vivid dream about a statue of a man with a body made of several different materials. This dream vexed him so, that he sought out someone to not only interpret the dream but to voice what the dream was in order to validate the interpretation. The wise men of renown were hesitant to voice the dream and so reluctant to do so that king Nebuchadnezzar threatened to kill all the wise men throughout his kingdom. When Daniel, a wise man, was advised of this he sought the mercies of God. God in a night vision revealed to Daniel Nebuchadnezzar's dream and its interpretation. For our purposes here, the primary elements of the dream of the statue are recorded in Daniel, Chapter 2, verses 24 through 45:

> *"Therefore Daniel went in unto Arioch, whom the king had ordained to destroy the wise men of Babylon: he went and said thus unto him; Destroy not the wise men of Babylon: bring me in before the king, and I will shew unto the king the interpretation (v.24) Then Arioch brought in Daniel before the king in haste, and said thus unto him, I have found a man of the captives of Judah, that will make known unto the king the interpretation. (v. 25) The king*

15 https://www.biblicalarchaeology.org/daily/biblical-artifacts/the-tel-dan-inscription-the-first-historical-evidence-of-the-king-david-bible-story/

answered and said to Daniel, whose name was Belteshazzar, Art thou able to make known unto me the dream which I have seen, and the interpretation thereof? (v. 26) Daniel answered in the presence of the king, and said, The secret which the king hath demanded cannot the wise men, the astrologers, the magicians, the soothsayers, shew unto the king; (v. 27) But there is a God in heaven that revealeth secrets, and maketh known to the king Nebuchadnezzar what shall be in the latter days. Thy dream, and the visions of thy head upon thy bed, are these; (v. 28) As for thee, O king, thy thoughts came into thy mind upon thy bed, what should come to pass hereafter: and he that revealeth secrets maketh known to thee what shall come to pass. (v. 29) But as for me, this secret is not revealed to me for any wisdom that I have more than any living, but for their sakes that shall make known the interpretation to the king, and that thou mightest know the thoughts of thy heart. (v. 30) Thou, O king, sawest, and behold a great image, the great image, whose brightness was excellent stood before thee; and the form thereof was terrible. (v. 31) This image's head was of fine gold, his breast and his arms of silver, his belly and his thighs of brass, (v. 32) His legs of iron, his feet of part iron and part of clay. (v. 33) Thou sawest till that a stone was cut out without hands, which smote the image upon the feet that were of iron and clay, and brake them unto pieces. (v. 34) Then was the iron, the clay, the brass, the silver, and the gold, broken to pieces together, and became like the chaff of the summer thresingfloors; and the wind carried them away, that no place was found for them: and the stone that smote the image became a great mountain, and filled the whole earth. (v. 35) This the dream; and we will tell the interpretation thereof before the king. (v. 36) Thou, O king, art a king of kings: for the

God of heaven hath given thee a kingdom, power, and strength, and glory. (v 37) And wheresoever the children of men dwell, the beasts of the filed and the fowls of the heaven hath he given into thine hand, and had made thee ruler over them all. Thou art this head of gold. (v. 38) And after thee shall arise another kingdom inferior to thee, and another third kingdom of brass, which shall bear rule over all the earth. (v. 39) And the fourth kingdom shall be strong as iron: forasmuch as iron breaketh in pieces and subdueth all things: and as iron that breaketh all these, shall it break in pieces and bruise. (v. 40) And whereas thous sawest the feet and toes, part of potters' clay, and part of iron, the kingdom shall be divided; but there shall be in it of the strength of the iron, forasmuch as thou sawest the iron mixed with miry clay. (v. 41) And as the toes of the feet were part of iron, and part of clay, so the kingdom shall be partly strong, and partly broken. (v. 42) And whereas thou sawest iron mixed with miry clay, they shall mingle themselves with the seed of men: but they shall not cleave one to another, even as iron is not mixed with clay. (v. 43) And in the days of these kings shall the God of heaven set up a kingdom, which shall never be destroyed: and the kingdom shall not be left to other people, but it shall break in pieces and consume all these kingdoms, and it shall stand for ever. (v. 44) Foreasmuch as thou sawest the stone was cut out of the mountain without hands, and that it brake in pieces the iron, the brass, the clay, the silver, and the gold; the great God hath make known to the king what shall come to pass hereafter: and the dream is certain, and the interpretation thereof sure. (v. 45)

Historically, what followed the rule of Babylon under Nebuchadnezzar was the Medo-Persian Empire (breast and arms [two arms – Mede / Persia] of sliver); the Greek Empire (belly and thighs of brass); the Roman Empire (legs [two legs] of iron, his feet part of iron and part of clay). The historical setting for the book of Daniel is around 604 BC (BCE) during the life of Nebuchadnezzar. The ascendancy of the Medo-Persian Empire was around 539 BC (BCE).The Greek Empire under Alexander the Great overthrew the Persian Empire in 334 BC (BCE). Upon Alexander the Great's death in 323 BC (BCE), the Greek Empire began to be divided up principally among his four generals, with two, Ptolemy I and Seleucus being the victors. Ptolemy I established the Ptolemy Empire and Seleucus established the Seleucid Empire. These two empires morphed into the Roman Empire which was initially seen as the western empire (Ptolemy) and eastern empire (Seleucid). According to most scholars, the Roman Empire was established in 27 BC (BCE) after Augustus Caesar declared himself emperor. In 285 AD (CE) emperor Diocletian divided the Roman Empire into the Western Roman Empire and the Eastern Roman Empire. The divided Roman Empire was unified under emperor Constantine and continued unified until his death in 337 AD (CE). His successor, Theodosius, divided the empire between his two sons Arcadius and Honorius, creating the Western Roman Empire and the Eastern Roman Empire. [16] According to historians, the fall of the Roman Empire dovetailed with the rise of Christianity and its advent is considered a factor in the fall of the Roman Empire.

- Babylonian (kingdom) -Nebuchadnezzar – head of gold
- Medo-Persian (kingdom) – breast and arms of silver
- Greek (kingdom) – belly and thighs of bronze
- Roman (kingdom) – legs of iron / feet part iron & part clay
- Christianity (Christ's kingdom) – consumes all the

---

16 https://www.rome.net/roman-empire

kingdoms and lasts forever

So, when this event from the book of Daniel is analyzed from both a historical and prophetical view, Daniel accurately explains the rise of kingdoms over the course of roughly 1,000 years. Such expanse of time and exactness serves to validate scripture.

In the New Testament of the Bible regarding Jesus of Nazareth, the book written by the apostle Matthew between 70 AD (CE) and 90 AD (CE)[17] in Chapter 27, verse 35 states, *"And they crucified him, and parted his garment, casting lots: that it might be fulfilled which was spoken by the prophet, The parted my garments among them, and upon my vesture did they cast lots."* Although the apostles Mark and John, and Luke wrote similarly of the event around a roughly the same time frame; the passages reference back to the Old Testament, Psalm 22:18 (*"They part my garment among them, and cast lots upon my vesture."*) which was written around 1044 BC (BCE).[18] The roughly 1,100 years between the initiation of the prophecy and the fulfillment of the prophecy lend credence to the validity of the Bible.

According the the Cambridge online dictionary, prophecy is defined as, " a statement that says what is going to happen in the future, especially one that is based on what you believe about a particular matter than existing facts." [19] Prophecy is relevant as validation of scripture in that the fulfillment of prophecy establishes as fact a belief that was prior held. Fulfillment of prophecy also dove-tails into scientific beliefs through

---

17 www.bibleodyssey.org/tools/ask-a-scholar/when-was-the-gospel-of-matthew-written.aspx
18 https://www.blueletterbible.org/study/parallel/paral18.cfm
19 https://dictionary.cambridge.org/us/dictionary/english/prophecy

mathematics. The mathematics of probability (the study of the likelihood of something happening) co-exists equally in analyzing fulfillment of scriptural prophecy or analyzing scientific observations of biology and cosmology.

Generally there are between 200 and 400 prophecies associated with Jesus of Nazareth, the Messiah, the Christ which have been fulfilled. [20] Of these are roughly 47 key prophecies fulfilled of general acceptance. Examples of just eight [21] are:

Messiah would be born in Bethlehem – Prophesied Micah 5:3 / *Fulfilled Matthew 2:1 (and Luke 2:4-6)*

Messiah would be born of a virgin - Prophesied Isiah 7:14 / *Fulfilled Matthew 1:22-23 (and Luke 1:26-31)*

Messiah would come from the line of Abraham – Prophesied Genesis 12:3, Genesis 22:18 / *Fulfilled Matthew 1:1 (and Romans 9:5)*

Messiah would come from the tribe of Judah – Prophesied Genesis 49:10 / *Fulfilled Luke 3:33 (and Hebrews 7:14)*

Messiah would be called Immanuel – Prophesied Isiah 7:14 / *Fulfilled Matthew 1:23*

Messiah would spend a season in Egypt – Prophesied Hosea 11:1 / *Fulfilled Matthew 2:14-5*

Messiah would be declared the Son of God – Prophesied Psalm 2:7 / *Fulfilled Matthew 3:16-17*

---

20 www.about-jesus.org/complete-chart-prophecies-jesus.htm
21 Author selected eight from a list of 47:
   https://www.learnreligions.com/prophecies-of-jesus-filfilled-700159

Messiah's bones would not be broken – Prophesied
Exodus 12:46 and Psalm 34:20 / *Fulfilled John 19:33-36*

The eight prophecies selected represent a breadth of issues
– political and religious, geographical locations and lineage
related to the Messiah in order to show the diversity of issues
relevant to prophecy and validation of scripture. Within
prophecy, the website Learn Religions put forth the following:

> "In the book <u>Science Speaks</u>, Peter Stoner and Robert
> Newman discuss the statistical improbability of one man,
> whether accidentally or deliberately, fulfilling just eight of
> the prophecies Jesus fulfilled. The chance of this
> happening, they say, is 1 in 10^{17} power. Stoner presents a
> scenario that illustrates the magnitude of such odds:
>
>> 'Suppose that we take 10^{17} silver dollars and lay
>> them on the face of Texas. They will cover all of
>> the state two feet deep. Now mark one of these
>> silver dollars and stir the whole mass thoroughly,
>> all over the state. Blindfold a man and tell him that
>> he can travel as far as he wishes, but he must pick
>> up one silver dollar and say that this is the right
>> one. What chance would he have of getting the
>> right one? Just the same chance that the prophets
>> would have had of writing these eight prophecies
>> and having them all come true in any one man,
>> from their day to the present time, providing they
>> wrote using their own wisdom.'" [22]

A similar analogy to further illustrate the improbability of
fulfillment of prophecy regarding the Messiah: Such fulfillment

---

22 https://www.learnreligions.com/prophecies-of-jesus-filfilled-700159

would be like asking a man to go out onto the Pacific Ocean and cast a fishing line to retrieve a ladies size five ring from the ocean on the first cast without knowing where and when to cast.

The improbability of fulfillment of scriptural prophecy based on the mathematics of probability enforces the validity of scripture.

Scripture has provability. History is contained within scripture. King David existed as a real historical figure. The ancient kingdoms of the Middle East existed in history. Jesus of Nazareth existed in history. The mathematics of probability is the same in prophecy just as it is in physics. Scripture has validity. With this in view let us delve into the timeline of creation.

CHAPTER 5
L.IGHT

Other than time, light might be one of the most mysterious components of creation. Without light though there is no time. Consider the biblical narrative in Genesis, chapter one, verses 2 through 5:

"And the earth was without form, and void; and darkness was upon the face of the deep. And the Spirit of God moved upon the face of the waters. (v.2) And God said, Let there be light: and there was light. (v. 3) And God saw the light, that it was good: and God divided the light from the darkness. (v. 4) And God called the light Day, and the darkness he called Night. And the evening and the morning were the first day. (v. 5)"

It was only upon creation of light and dividing the light from the darkness that time began. Evening and morning being components of light and the day before the darkness and night and such was day one: the beginning of time. But, what is light?

Light is a form of electromagnetic radiation that enables the human eye to see or make thing visible. It is also defined as radiation that is visible to the human eye. Light contains photons, which are minute packets of energy.[23] John Hopkins University states, "...Visible light is a tiny portion of a huge smorgasboard [sic] of light called the _electromagnetic spectrum_. For our convenience, we break this smorgasboard [sic] up into different courses (appetizer, salad, etc.) and refer to them by

---

23 https://byjus.com/physics/light-energy/

name, such as gamma-rays, X-rays, ultraviolet, optical, infrared, and radio. However, it is important to remember that they are all just *light*. There are no 'breaks' and no hard boundaries in the electromagnetic spectrum—just a continuous range of energy."[24]

A photon is defined as: a type of elementary particle which has a zero rest mass and moves with a speed of light in the vacuum. "A photon is the 'quantum of electromagnetic radiation'. In other words, it is the smallest and the fundamental particle of an electromagnetic radiation. A photon has no mass, no electric charge and it is a stable particle. These particles possess wave-particle duality. It has 2 polarization states. Light, for example, is an electromagnetic radiation. Photons can be said to be the fundamental particles of light. It can be said that light is carried over space by photons."[25]

In the above quote regarding the photon, notice that a photon is considered a fundamental particle of light but also that light is carried over space by photons. This explanation does not provide any clarity as to the nature of light. In fact, it devolves to circular reasoning as to the photon-light – light-photon relationship. An additional confusion is that light is defined as electromagnetic radiation yet, a photon has no observed electric charge. An electric charge such as an electron with a observed negative charge or a proton with an observed positive charge lie within the realm of electromagnetic definition – but, a particle with no electric charge is considered as being electromagnetic radiation?

Despite the fact that light is identified as a particle – a photon – experimentally and observationally it exhibits

24 https://blair.pha.jhu.edu/spectroscopy/basics.html
25 https://byjus.com/physics/electrons-and-photons/ (NOTE: quoted text is just a partial portion of the definition contained at the quote source)

properties of and is transmitted across distances as a wave. Accordingly, two integral elements of light are wavelength and frequency. The frequency of the wave is the time for to peak of the wave to pass by a certain point within a given amount of time. The wavelength is the distance between peaks. There is a specific frequency and specific wavelength for each color of light and in fact for each type of electromagnetic radiation. These specific frequencies and wavelengths are mathematically quantified using a the equation: $c=f\lambda$ wherein "c" is the assumed speed of light constant , "f " is the frequency and "λ" is the wavelength. The equation stated without the use of symbols is: the speed of light is equal to the frequency multiplied by the wavelength.

The speed of light in scientific literature is 186,282 miles per second. This is the speed in a vacuum as calculated by experimentation. Space.com states, "According to physicist Albert Einstein's theory of special relativity, on which much of modern physics is based, nothing in the universe can travel faster than light. The theory states that as matter approaches the speed of light, that matter's mass becomes infinite. That means the speed of light functions as a speed limit on the whole universe. The speed of light is so immutable that, according to the U.S. National Institute of Standards and Technology, it is used to define international standard measurements like the meter (and by extension, the mile, the foot, and the inch). Through some crafty equations, it also helps define the kilogram and the Kelvin."[26] To non-scientists this standard is accepted without question.

When considering Einstein and the speed of light, his famous equation: $E=mc^2$ requires evaluation. The equation stated without the use of symbols is: Energy equals mass

26 https://www.space.com/15830-light-speed.html

multiplied by the squared speed of light. The outcome of the equation formulates the theoretic creation of energy (the "E" in the equation). It is this equation ($E=mc^2$) that was the basis for the theory and development of the atomic bomb. The atomic bomb being a bomb that created pure energy from matter and changed the course of World War II during the 1940's and the future of weapons of war. When evaluating the equation $E=mc^2$ there are three components "E" being energy, "m" being mass and "c" being the speed of light. However, note that in the equation "c" is to the second power commonly referred to as "squared". Mathematically, squaring a component is multiplying the component by itself. In theory the speed of light could be multiplied by itself. The creation of the atomic bomb was scientific experimental validation of the $E=mc^2$ theory. Thus, with the theory validated; the speed of light is known to be at a minimum the speed of light multiplied by itself – at least during the creation of energy. So, instead of traveling at 186,282 miles per second (as generally referenced) it travels at 34,700,983,524 miles squared per second squared [186,282 mi/s X 186,282 mi/s]. 34.7 billion miles squared per second squared.

Although the nomenclature of miles squared and seconds squared seems convoluted, mathematical formulation requires that upon the squaring of a value, the units of the value are also required to squared. Therefore, with the accepted standard speed of light being 186,282 miles per second: the squaring of 'c' (the standard speed of light) is $186,282^2$ miles^2 per second^2.

Within physics there are two main approaches to explaining the squaring of 'c' – because there is nothing faster than the speed of light. The first approach is that 'c' is a "mathematical constant" (the speed of light) and therefore it is the "mathematical constant" that is squared and not the respective value of the "mathematical constant". The second approach is

38

that the "mathematical constant" is squared because that is what is required to make the math work. However, both of these approaches are nonsensical in that squared units are appropriately accepted and applied in other calculations related to energy and motion.

Regarding the speed of light, the squaring of miles indicates a dimensional component of area: miles by miles. This can be conceptualized by visualizing a square mile of land as being one mile long by one mile wide.

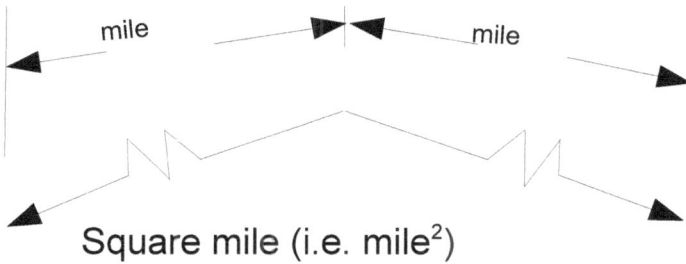

Square mile (i.e. mile^2)

However, since light travels as a wave form, the dimensionality of peaks and troughs of a wave is present. This waveform is confined in area due to characteristics related to its wavelength. This wavelength area is measured for visible light between 400 nanometers for violet light and, at the opposite end of the spectrum, 750 nanometers for red light. A nanometer is one-billionth of a meter or 39 hundred-millionths of an inch (.000000039 [3.9 x 10^{-8} inches]). Thus the confining width of the waveform elongates the length to conform to the mathematical squaring.

39

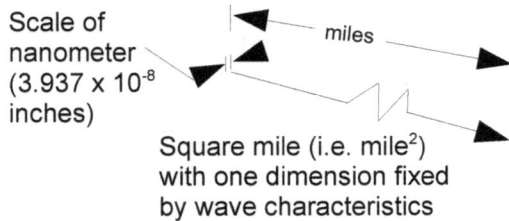

Scale of nanometer (3.937×10^{-8} inches)

miles

Square mile (i.e. mile²) with one dimension fixed by wave characteristics

So, instead of light traveling at 186,282 miles per second wide by 186,282 miles per second long: it travels at waveform width – for visible light between 15 hundred-thousands of an inch (violet light) and 29 hundred-thousands of an inch (red light) – and a length (distance) of between 8.781 quadrillion [8.781×10^{15}] miles per second (red light) and 4.786 trillion [7.786×10^{12}] miles per second (violet light). And, the squaring of seconds indicates a continuity of time. Light travels at a speed in miles per second from its creation and for each successive second of time thereafter. It is important to note that the "speed of light" was not altered in the calculation of these distances. The only change is recognition of the dimensional component related to wavelength characteristics.

The relevance of this energy equivalency formula – $E=mc^2$ – is that the multitude suns of the multitude of the galaxies within the universe are emitting energies created in the manner of atomic bombs. And these energies are in a state of constant creation. They are, after all, the starlight we see when observing the night sky.

While the current "scientific accepted" absolute speed of light of 186,282 miles per second is generally perceived as incontrovertible, is it. Experimentally, the speed of light as measured is for the two-way distance being measured and not directly as a one-way measurement. As such, there is the theoretical possibility that the one way speed of light is

40

significantly different. According to the website that explains and challenges engineering and science, Veritasium - the speed of light is potentially so significantly different that it is instantaneous and observable at the exact instance it is created.[27]

Contemplation of the squared speed of light and distances as discussed above gives pause. Is a distance of 4 trillion miles in one second instantaneous? Is a distance of 8 quadrillion miles in one second instantaneous?

From the current scientific point of view, the electromagnetic field is the fundamental quantum field. It is from this field that light emanates and provides the basis from which all other quantum fields and quantum particles exist and interact: "The electromagnetic vacuum field is not just a background entity but a fundamental component of quantum-mechanical ontology. Its dynamic nature and ubiquitous presence influence every aspect of quantum systems, from atomic transitions to the structure of spacetime."[28]   As the fundamental quantum field its creation and existence as the first object of creation is a necessity. It was the first spoken aspect of creation by God, *"And God said let there be light: and there was light. (v. 3) And God saw the light, that it was good:...(v. 4) "*[29]

---

27 https://youtu.be/pTn6Ewhb27k

28 https://www.hilarispublisher.com/open-access/the-electromagnetic-
   vacuum-field-a-fundamental-component-of-quantummechanical-
   ontology.pdf   : Ru, Wang. "The Electromagnetic Vacuum Field: A
   Fundamental Component of Quantum-mechanical Ontology." J Phys Math
   15 (2024): 510

29 Genesis 1:3-4

Thomas W. Guthrie

CHAPTER 6
E.ARTH

"In the beginning God created the heaven and the earth. (v. 1)And the earth was without form, and void; and darkness was upon the face of the deep. And the Spirit of God moved upon the face of the waters. (v. 2)"...

And God said, Let there be a firmament in the midst of the waters, and let it divide the waters from thew waters. (v. 6) And God made the firmament, and divided the waters which were under the firmament from the waters which were above the firmament: and it was so. (. 7) And God called the firmament Heaven. And the evening and the morning were the second day. (v. 8) And God said, Let the waters under the heaven be gathered together unto one place, and let the dry land appear: and it was so. (v. 9) And God called the dry land Earth; and the gathering together of the waters called he Seas: and God saw that it was good. (v. 10)..." Genesis 1:1-10

In these verses note that 'the waters' are the entirety of the underlying component from which the physical universe was created. Then 'the waters' were then divided by 'the firmament'. So, after creation of the firmament there were 'the waters' <u>above</u> the firmament and were 'the waters' <u>under</u> the firmament. From this point forward in scripture there is no additional explanation as to what happened to 'the waters' above the firmament. But, there is detailed explanation regarding the firmament and as the waters that were divided under the firmament.

This detailed explanation of the waters divided under the firmament are addressed in verses nine and ten of Chapter 1 of the book of Genesis:

> *"And God said, Let the waters under the heaven be gathered together unto one place, and let the dry land appear: and it was so. (v. 9) And God called the dry land Earth; and the gathering together of the waters called he Seas: and God saw that it was good. (v. 10)"*

The geological sciences affirm the Genesis creation account of the earth. There was water and the ground appeared from the water. Observable evidence throughout the earth shows the predominance of the sedimentary nature of the earth. River gorges throughout the world cutting through layers and layers of different sediment show this nature. The uplifted mountains of the Himalayas, the Altay, the Kunlun, the Zagros and the Ural on the Asian continent; the Rwenzori and the Drakensberg on the African continent; the Andes on the South American continent; the Alps on the European continent; the Adirondacks, the Great Smokey, the Cascades and the Rockies on the North American continent all show the sedimentary nature of the earth. These named mountain ranges are but a few of the overall mountain ranges of the earth. And, except for a few mountain ranges derived from volcanic activity; the vast majority of mountains in the world are sedimentary in nature.

Even in man's endeavor for knowledge and wealth, the earth shows the evidence of the sedimentary nature. The drilling of water wells into the earth, the drilling of oil and gas wells into the earth and mining activities all show the sedimentary nature of earth.

The elements of oxygen, silicon, aluminum, iron, calcium

sodium, potassium and magnesium make up 98.4% of the earth's crust. All the other elements constitute the remaining 1.6% of the earth's crust. [30] Currently, there are 118 identified elements.

Contextually within Genesis, the waters is consistent in use from the Spirit of God moving upon the face of the waters in verse two through the gathering together of the waters on earth to create the seas in verse 10. Such usage would indicate the physical attribute of water. However, could it be conjectural to imply 'the waters' is in reference to underlying elements of water – hydrogen and oxygen? Current scientific understanding opines that hydrogen is the most abundant element in the universe and oxygen is the third most abundant element in the universe. When hydrogen and oxygen are combined in a stable molecule of two hydrogen atoms and one oxygen atom the resultant molecule is water – H_2O.

Current scientific theory regarding the evolution or synthesis of elements uses the hydrogen atom as the basic building block. This synthesis of elements involves the fusion of protons in a nuclear reaction such as within a star. General education regarding the atom teaches that an atom consists of a nucleus containing a proton and a neutron, and an electron or electrons that 'orbit' around the nucleus. It the the number of protons that differentiate each element. In the Periodic Table[31], the number of protons are shown as the atomic number for each element.

For the eight elements that make up 98.4% of the earth's crust (oxygen, silicon, aluminum, iron, calcium sodium,

---

30 https://www.thoughtco.com/chemical-composition-of-earths-crust-elements-607576

31 An organized chart of each element showing, atomic number, atomic weight, element abbreviation and grouping(s).

potassium and magnesium), the atomic number for each is as follows: oxygen-8, silicon-14, aluminum-13, iron-26, calcium-20, sodium-11, potassium-19 and magnesium-12.

According to Britannica online, "In 1938 German-born physicist Hans Bethe proposed the first satisfactory theory of stellar energy generation based on the fusion of protons to form helium and heaver elements. He showed that once elements as heavy as carbon had been formed, a cycle of nuclear reaction could produce even heavier elements."[32] Carbon has an atomic number of 6. And, importantly carbon is the basic element necessary for all lifeforms on earth.

The theory of stellar element synthesis holds merit. However, accord to the creation account of Genesis the sun and stars were created on the fourth day of creation AFTER the creation of earth on the third day of creation.[33] So, the earth and all of the elements of the earth (carbon, oxygen, silicon, aluminum, etc.) came into existence prior to any stellar activity.

Other than "God said...", an alternate theory of element synthesis is possible. Element synthesis involves nuclear reactions. Such reactions existing within the core of the earth itself are plausible. Geophysicists have long opined that the core, the innermost geological layer, of the earth is a very hot, very dense ball at the center of the earth. Current geological theory is that the heat of the core is related to, "(1) heat from when the planet formed and accreted, which has not yet been lost; (2) frictional heating, caused by denser core material sinking to the center of the planet; and (3) heat from the decay of radioactive

---

32 https://www.britannica.com/science/physical-science/Evolution-of-stars-and-formation-of-chemical-elements

33 Genesis 1:9-19

elements."[34] A nuclear generator at the center of the earth would also explain a dense hot core. And with the understanding of the nuclear fusion reaction using the deuterium and tritium isotopes[35] of hydrogen the abundance of hydrogen fuel as the gathering of the waters under the firmament would be boundless.

Planetary accretion is a scientific theory that dust, gas and solid materials, at some point, under some condition coalescence into a body. Of course though dust, gas and solid materials would have to first coalesced in their respective forms from the elements created by fusion reactions in the stars. After the initial accretion into a planetary body, questions of formation processes then arise. Why are there (as theorized) four basic components of earth: core, mantle, crust and atmosphere? How are these four basic components separated? Why is the crust stratified as evidenced in observable sedimentary formations? How was the crust stratified? Although gravity and centrifugal force of a rotating body are likely contributing factors, how much more complex are the processes? When considering the Genesis story of creation, the complexity of scientific planetary accretion theory is exacerbated as it is stated that the seas, the dry land and vegetation all appeared on the third day of creation:

> *"And God said, Let the waters under the heaven be gathered together unto one place, and let the dry land appear and it was so. (v.9) And God called the dry land Earth; and the gathering together of the waters called he Seas: and God saw that it was good. (v.10) And God said, Let the earth bring forth grass, the herb yielding fruit*

---

34 https://www.scientificamerican.com/article/why-is-the-earths-core-so/
35 An isotope of an atom contains additional neutrons in the nucleus. The base hydrogen atom has one neutron. Deuterium has two neutrons and tritium has three neutrons.

after his kind, whose seed is in itself, upon the earth: and it was so. (v. 11) And the earth brought forth grass, and herb yielding seed after his kind, and the tree yielding fruit, whose seed was in itself, after his kind: and God saw that it was good. (v. 12) And the evening and the morning were the third day (v.13) "[36]

Thus, according to Genesis not only were the elements beyond hydrogen and oxygen created, the elements coalesced into dust, gas and solid materials and accreted into a planetary body: but, the dry land and the seas were separated, and life came into existence all on the same day.

There is observable evidence of life existing throughout the the stratification of sedimentary layers also known as the geologic column, The science of Geology groups and labels different layers into geologic time frames. From the deepest layers (classified oldest) to the top layers (classified youngest) these strata are labeled: Precambrian, Cambrian, Ordovician, Silurian, Devonian, Carboniferous, Permian, Triassic, Jurassic, Cretaceous, Paleogene, Neogene and Quaternary. Of note, rocks deeper than Precambrian are classified as "basement rock" and are non sedimentary in composition. The hypothesis underlying the geologic column and the distribution of fossils within is a continuing cycle of life, death of life, deposition of sediment over hundred of millions of years. The Precambrian period is hypothesized to have begun around 650 million years ago.

The Grand Canyon located in the State of Arizona in the United States is one of the most studied observable examples of the geologic column. At 6,000 feet of depth, over one mile in depth it is remarkable. However, oil wells have been drilled at over five miles in depth. Whether studying the geologic column

---

36 Genesis 1:9-13

of the Grand Canyon or the geologic column within well bores there is surprising uniformity. Although there are some local and regional variances, or unconformities; it is the uniformity that allows the science of Geology to classify and label the geologic column. Sediment, according to scientific theory is the erosion and deposition of soil and decomposing plants and animals. Yet, where did the original soil come from to erode? And, where did the original soil erode from and where did it get deposited?

The predominate geological theory of earth's history on one of slow, incremental, evolutionary changes over long periods of time. However, an alternate theory of cataclysmic – sudden, violent and unusual - changes over short periods of time does exist and has observable evidence to substantiate it. This alternate theory is known as catastrophism.

Catastrophism includes such events as earthquakes, meteorites, volcanic eruptions and tsunamis. From a biblical based perspective, the geological theory of catastrophism can be supported:

> *"The words of Amos. two years before the earthquake"* [37]

> *"...rained upon Sodom and upon Gomorrah...brimstone and fire...out of heaven."* [38]

Regarding the stratification of sediment, biblically there are two avenues for the stratification; initial creation of the earth and the flood recorded in the days of Noah. Both events would fit catastrophism, sudden and unusual. Creation can be recognized as such, but, what about the flood of Noah?

Traditionally, the flood of Noah is only recognized for the

---

37 Amos 1:1
38 Genesis 19:24

forty days and nights of rain. [39] However, the waters flooding the world were not only from rain but, the interior of the earth:

> *"In the six hundredth year of Noah's life, in the second month, the seventeenth day of the month, the same day were all the fountains of the great deep broken up, and the windows of heaven were opened."* [40]

From experience, man has throughout history known water existed not only above ground in seas, lakes, rivers and streams but, also under the ground. And he has exploited technological accessible underground water. But, a yet more impactful understanding of the 'fountains of the great deep' has been brought to light by scientists. An online article by Andy Coghlan in New Scientist dated June 12, 2014 titled "Massive 'ocean' discovered towards Earth's core" [41], states that scientists have discovered a reservoir of water containing three times the volume of all the oceans in a blue rock called ringwoodite located in the earth's mantle about 700 kilometers below the surface of the earth,

Underground sources of water are many times under pressure and when released to the surface become powerful geysers. Understanding this concept it can be seen that the fountains of the great deep when broken open might have created great geysers. And, when geysers erupt – especially violently – sediments are transported through the geyser stream to be ultimately deposited or re-deposited.

Probing a logical course of thought; with the flood of Noah,

39 Genesis 7:4
40 Genesis 7:11
41 https://www.newscientist.com/article/dn25723-massive-ocean-discovered-towards-earths-core/

sediments transported by geysers would have remained as suspended solids in the turmoil of the rain and the waters from the deep. As the turmoil of the event subsided, sediments would filter, settle, stratify and compact. Along with the mechanical aspects of stratification there would also be chemical aspects resulting in cementing and other chemical bonding of soils. As an example, concrete is made from the mixing of sand, gravel, limestone, marl, clay and water. The limestone, marl and clay are the additives which provide the underlying base elements of calcium, silicon, aluminum and iron.

The length of time for rock formation is also not problematic under this scenario. In the example of concrete it can support a load of heavy machinery in seven days and be fully cured in 28 days. With the addition of calcium carbonate the curing time for concrete can be accelerated substantially. Diamonds thought to be formed miles under the surface of the earth at high pressure and temperature over billions of years are now being produced in a lab at room temperature in 15 minutes with a mixture of gallium, nickel, iron and silicon.[42]

The plants and animals killed during the flood would settle in manners and processes consistent with gravity, permeability and hydraulic concentration. This of course would result in the sedimentary strata and fossil distribution as observed. And, since the various life forms were likely grouped in survival units or herds there would not be an overall uniformity of distribution within the fossil record. Strata deposited during the flood would be additions to existing strata of the earth resultant of original creation. Subsequent draining of the flood waters from the

42 https://www.snexplores.org/article/lab-diamonds-without-extreme-pressure#:~:text=The%20process%20starts%20with%20a%20liquid%20of%20gallium%2C,lock%20together%20in%20a%20diamond%20's%20crystal%20structure.

earth's surface would result in evidenced large scale erosion.

Such a logical course of thought is justifiable. Yet, as one's observation of the earth expands, a new question arises: Why are fossils, especially those of sea creatures found on mountain tops?

According to current predominate geological theory, mountain formation was primarily the result of continental fusion due crustal plate tectonics. Observation of the continents of the global earth seem to indicate that at one time in history all the continents fitted together as one giant single continent. The concept of a single giant continent was first scientifically introduced in 1912 by German meteorologist Alfred Wegener and named the single continent Pangea. Later geological studies indicate that the super continent named Pangea was superseded in geological time by two massive continents named Laurasia consisting in the northern part of the global earth and Gondwana consisting in the southern part of the global earth. Subsequent to the fusion of Laurasia and Gondwana to form Pangea, the continents separated into the continental configuration currently observed via continental drift also attributed to plate tectonics.

From a scientific perspective this creation, fusion and separation of continents is palatable. But, how does such a scenario fit into a theological perspective?

Theologically, the world wide flood of Noah is a given and affirmed within scripture. And, although subtle, such separation is identified in scripture. In the book of Genesis, chapter 11, verse one it is stated that, *"All the whole earth was of one language, and of one speech."* This chapter of Genesis is the account of the peoples of the earth subsequent to the flood of Noah and repopulation of the earth coming together in the land of Shinar to build a city and a tower to heaven, commonly referred to as 'the tower of Babel'. In this account, the peoples came, *"...lest we be scattered abroad upon the face of the whole earth."*[43] Such activity by the peopled displeased the Lord. Later in the account, *"So the Lord scattered them abroad from thence upon the face of all the earth: and they left off to build the city."*[44] When contemplating this account and these verses; why were the people concerned about being scattered abroad upon the face of the whole earth and by what means did the Lord scatter the peoples abroad upon the face of the earth? Was it by supernatural 'translation' whereby the Lord placed different peoples at different locations upon the earth: or, did the Lord use the agency of his creation as the mechanism for the scattering?

The answer to the questions proffered above likely lies in Genesis chapter 10 with the lineage of the sons of Shem, one of the sons of Noah. In verse 25 of chapter 10 of Genesis, *"And unto Eber were born two sons: the name of one was Peleg; for in his days was the earth divided..."* The name Peleg literally means "to split". The author[45] of **Secrets of the Ancients** identified that it was at the time of Peleg that the continental separation with the continents configured as presently observed began to take place. Within the biblical lineage, Peleg was the fourth generation after the flood. Shem the son of Noah begat Arphaxad; Arphaxad begat Salah; Salah begat Eber; Eber begat Pelag. In the time line of creation, Arphaxad was born two years after the flood of Noah, Salah was born 35 years later, Eber was born 30 years later and Peleg was born 34 years later.[46] So, the time from the flood to the birth of Peleg was 101 years.

---

43 Genesis 11:4
44 Genesis 11:8
45 The exact author hasn't been validated for this cite as this point was synthesized years ago and there are several books Secrets of the Ancients.
46 Genesis 11:10-16

Returning to the narrative of the tower of Babel in chapter 11 of Genesis, reflect again to the actions of the peoples in their assembly in the land of Shinar. They assembled, *"lest we be scattered abroad upon the face of the whole earth."* Contemplate what events that may have been taking place for the peoples to surmise that they may be scattered abroad upon the face of the whole earth. The likelihood is that catastrophic activity was taking place that was recognized by the people and by coming together, there was an attempt to circumvent their dispersal across the face of the earth.

Such speculation may initially seem absurd but, there is archaeological evidence across the globe indicating that many people groups experienced a concern about earth events. It can be postulated that these concerns led to the creation of structures to track the stars; as there is a predominance of scientific opinion that heavenly events were central to ancient civilizations. One example is the archaeological site of Gobekli Tepe located in the country of Turkey. Gobkli Tepe is purported to be possibly the world's first observatory. However, there are also other archaeological structures related to tracking the stars such as the famed Stonehenge structure located in the country of Great Britain. Although these structures may vary in the time of their construction, their purpose is aligned. Why was tracking the heavens so important? Was it just for agriculture planting cycles as proposed by many archaeologists?

Major geological catastrophism is generally relegated by scientists to time periods perceived to be devoid of human presence. Yet, the evidence of cataclysmic events is observable throughout the earth. There are several places where meteoric impacts are evidenced. One of the most observable is the meteor impact crater named the Barringer Meteor Crater located in the State of Arizona in the United States of America. Although relatively small in size at ¾ miles wide and an estimated 600 feet deep, the impact would have annihilated a city the size of the current size of Kansas City, Missouri[47] In 2020 the size of Kansas City, Missouri was roughly 312 square miles in size (17.66 miles x 17.66 miles) with a population of just over 508,000. [48] While this impact site is relatively insignificant in relationship to the size of the earth, if impacts were occurring across the earth at the same time the overall significance increases exponentially.

Unlike the relatively small meteor creating the Barringer Meteor Crater in Arizona: the Chicxulub meteor (asteroid) impact near the Yucatan Peninsula of the country of Mexico was a world changing event. At an estimated diameter of 6.2 miles, it left an impact crater of an estimated 110 mile diameter with a 12 mile depth. The outcome of this impact is theorized to have created whole earth impacting earthquakes, tsunamis, volcanic eruptions, wildfires, vaporized water and rock while also expelling debris into the atmosphere creating a world-wide darkness and temperature cool down. It is this asteroid impact identified by scientists as the cause of the massive and species extinction of the dinosaurs. Recently, reported in 2022, a newly found meteor impact crater was located in the ocean off the West coast of Africa. This newly identified impact crater is named the Nadir crater. Research indicates that the impact occurred around the same time frame as the Chicxulub impact.[49] The Nadir crater like the Barringer Meteor Crater is relatively small in size at an estimated ¼ mile diameter. The underlying focus though is that the Nadir crater was created in a time frame with the Chicxulub

47 https://earthobservatory.nasa.gov/images/148384/arizonas-meteor-crater
48 https://en.wikipedia.org/wiki/Kansas_City,_Missouri

49 https://www.sciencealert.com/a-mysterious-crater-found-in-the-ocean-may-be-a-new-clue-to-the-dinosaurs-fate

crater. As opined previously, if impacts were occurring across the earth at the same time the overall significance increases exponentially.

In addition to asteroid and meteor impacts there has been speculation the planet sized celestial bodies have interacted with the earth. Scientific speculation surmises that a planetary body the size of mars impacted the earth and this impact caused the creation of earth's moon.[50] There are points of view derived from historical records that there have been other planetary sized bodies interacting with earth.

In regard to world inhabitants being concerned about being scattered across the face of the earth, the probability exists that the heavens were being watched closely because there was evidence in the heavens warranting such concern. And, God through the agency of creation used asteroid impacts and planetary fly-bys with an ultimate outcome of continental separation bringing forth the people being scattered across the face of the earth.

Physical observations of the earth shows that the lands were created from water, water reformed the lands and the lands were rearranged. Biblical references in Genesis show the lands were created from water, water reformed the lands and the lands were rearranged.

50 https://www.livescience.com/64572-planetary-collision-life-earth.html

CHAPTER 7
M.AN

In the context of creation, man is an enigma. From a secular scientific viewpoint theory 'man' transitioned over time from other species to the form and the being constituted as presently exists as human. This 'evolution' of 'man' is theorized to evolve at the point of humanoid likeness from ape-like species through seven specific transitions [listed oldest to most recent]: Australopithecus, Homo Habilis, Homo Erectus, Homo Neanderthalensis (a/k/a Neandethal), Homo Sapiens, Cromagnon. Modern man. Yet, from a traditional biblical viewpoint 'man' is; singularly created, a single, independent and distinct being. Man is always as one: Adam.

Although the theory of evolution in its broadest sense is contentious between the scientific community and the theological community, in its narrowest sense the theory of the evolution of man is a singular point of contention without apparent resolution. However, as scientific inquiry advances and additional evidence comes to light, the theory of evolution of man comes under reduction. No longer is there a transition of man from one 'type' of human to another but, there are different 'branches' of man. A key example of this reductionism is illustrated with studies of Neanderthal man. Marianne Guenot, a science writer in Business Insider in an article dated January 1, 2024 writes, in part, "...Recent discoveries, however, are upending that view [Neanderthal losing the evolutionary battle against Homo sapiens] and reigniting a debate among scientists about whether Neanderthals should be considered to be the same species as early modern humans."[51] The article further states, "That view started to change in 2008 when the Swedish geneticist Svante Pääbo...sequenced the genome of a Neanderthal by extracting DNA from ancient bones...Through his research...he showed most living humans carried about 2% of Neanderthal DNA."[52] Writer Tia Ghose in an article in Live Science dated February 7. 2024 writes, "In the early 1900s, scientists conceived of Neanderthals as apelike and almost bestial. But in the past few decades, unambiguous evidence has indicated that our closest human relatives mated with us at multiple points in time..."[53] This excerpt is based on an interview with archaeologist Ludovic Slimak related to his book, "The Naked Neanderthal: An New Understanding of the Human Creature" (Pegasus Books, 2024). Mr. Slimak makes an enlightening observation as reported by Ms. Ghose, "...we all have a certain degree of Neanderthal DNA...When you are searching for ancient DNA...all these early sapiens have recent Neanderthal DNA...But when you reach and try to extract DNA from the last Neanderthals...contemporaries of these early sapiens...there's not a single Neanderthal with sapiens DNA."[54]

From a theological perspective: *"And God said, 'Let us make man in our image, after our likeness; and let them have dominion'...And the evening and the morning were the sixth day."*[55] This passage taken from Genesis 1: 26, 31 sets the basis for most Christian teaching about the creation of Adam. It is taught that Adam (a/k/a man) was created on the sixth day of

---

51 https://www.businessinsider.com/neandethals-humans-belong-to-same-species-could-rewrite-history-evolution-2023-12
52 Ibid
53 https://www.livescience.com/archaeology/simply-did-not-work-mating-between-neandethals-and-modern-humans-may-have-been-a-product-of-failed-alliances-says-archaeologist-ludovic-slimak
54 Ibid
55 Genesis 1: 26, 31

creation. This idea seems fairly straight forward and is appealing to a basic grasp for biblical truths. However, upon studying the creation of Adam in depth, the premise of Adam being created on the sixth day clearly becomes invalid.

How then is it determined when Adam was created? Does the Bible really say? After all, the Bible says point blank that it was the sixth day when man was created, doesn't it? The detailed creation of Adam is contained in Genesis 2: 4-25. When adapting the information contained in these verses to the chronology and context of other verses in Genesis it can be seen when Adam was actually created.

When God created Adam from the dust of the ground as recorded in Genesis 2:7, the timing of the creation of Adam can clearly and intelligently be followed. Starting in Genesis 2: 4 - *"These are the generations of the heavens and of the earth when they were created, in the day that the Lord God made the earth and the heavens."* Traditionally, this verse is associated with verse three and is seen as a closing clause to the seven days of creation. However, this verse is an introductory clause and not a closing clause. It introduces the generations (things going on in creation) when the heavens and the earth were created. An identical introductory clause is seen in Genesis 5:1 when the generations of Adam are introduced. Verses four through 25 of Genesis Chapter two are a parenthetical statement within the narrative of creation detailing the creation of Adam with attention to day three of creation and day six of creation. Specifically however, the creation of Adam is associated to the last half of Genesis 2:4, *"...in the day that the Lord God made the earth and the heavens,"* It is with this phrase that the creation of Adam is introduced.

From the introductory clause of the second half of Genesis

2:4 it can be seen that the text following it is in relation to the time when the heavens and the earth were created. When though were the heavens and earth created? Genesis 1, verses 9 provides the answer:

> *"And God said, Let the waters under the heaven be gathered together unto one place, and let the dry land appear: and it was so. (v. 9) And God called the dry land Earth; and the gathering together of the waters called the Seas: and God saw that it was good. (v. 10) And God said, Let the earth bring forth grass, the herb yielding fruit after his kind, whose seed is in itself, upon the earth: and it was so. (v. 11) And the earth brought forth grass and herb yielding fruit after his kind, whose seed is in itself, after his kind: and God saw that it was good. (v. 12) And the evening and the morning were the third day. (v. 13)"*

Although heaven was created on the second day, earth and its vegetation were specifically created on the third day. Thus, in the day heaven and earth were created as proscribed in Genesis 2:4; the focal point is the third day. Therefore, Adam, in the day that God made the earth and the heavens, the third day, was formed from the dust of the ground, Adam was created on the third day.

Notice also that the earth yielded forth its fruit on the third day. The creation of vegetation on the third day as recorded in Genesis 1:11-12, *"And God said, Let the earth bring forth grass, the herb yielding fruit after his kind, whose seed is in itself, upon the earth: and it was so. (v. 11) And the earth brought forth grass and herb yielding fruit after his kind, whose seed is in itself, after his kind: and God saw that it was good. (v. 12)"* is entirely consistent with the chronology expounded in Genesis

2:5-9, *"... And the Lord God formed man of the dust of the ground...(v. 7)... And out of the ground made the Lord God to grow every tree that is pleasant to the sight, and good for food; (v. 9)..."*

The creation of vegetation seems to be a stumbling block in most teachings. Traditionalists holding that Adam was created on the sixth day finely divide the expressions contained in Genesis 2: 5-9. They say that keys to the entire creation of Adam are related to the phrases "herb of the field" (Genesis 2:5) and "And the LORD God planted a garden" (Genesis 2:8). However, the reason there was no herb of the field is given in Genesis 2:5. There was no herb of the field because 1) God had not caused it to rain and, 2) no man existed to till the earth.

God shows that Adam was created from the "dust of the ground". The language of dust is entirely consistent with the idea that God had not caused it to rain. For God did not say that Adam was created from the "mud of the earth". While this analogy appears flippant, the reality of the situation shows the timing of the creation of Adam. Without rain and without man, the herb of the field did not exist. But the herb of the field is not just related to vegetation cultivated by man, it also in context refers to vegetation in general. While it is true that there was no herb of the field prior to the creation of Adam it can also be seen that there was no other vegetation either.

When critically reviewing Genesis 2:8-9 it can been seen that the trees created for food and sight aren't just for the garden but in context of creation itself. Consider, for God to have planted a garden in Eden he would have had to create the trees before planting the garden or created the vegetation at the same time as he planted the garden. Either situation places the creation of the garden within the same chronology that exists in Genesis

1:11-13, *"...And the evening and the morning were the third day. (v. 13)"*

Following the chronology of the creation of vegetation as it relates to Adam, it is seen that Adam (Genesis 2:7) was created before the creation of vegetation described in Genesis 2:8-9. And since vegetation was created on the third day of creation, Adam was created on or before the third day of creation.

What about the sixth day though? The Bible clearly states in reference to the sixth day, *"And God said, Let us make man in our image, after our likeness:..."* (Genesis 1:26). Continuing the study of Adam from the chronology contained in chapter two of Genesis it can be seen that the Bible is indeed true and correct, but not from the perspective from which it is traditionally viewed regarding the creation of Adam. Although translations of the Bible use the term "man" in Genesis to indicate both man in the singular (a/k/a Adam) and man in the plural (a/k/a Adam and Eve), only by following the chronology detailed in the Bible contextually can the two be differentiated.

In Genesis 1:26 when God say, *"Lets us make man in our image,..."* he later in the same verse utilizes the phrase, *"...and let them..."* Now, since "them" is grammatically the plural of man, it must considered to whom does "them" refer? An important note to this question is that "them" in Genesis 1:26 is stated prior to any introduction of the concept of male and female. Previously, it has been shown that Adam was created on the third day. The reference to man in the plurality occurs on the sixth day. Is it likely that since Adam was created on the third day the reference to man in the plural (them) is strictly associated with the creation of Eve; or even a possibility of a subsequent creative event?

Clearly understand that Genesis 2:4 introduces the generations of the heavens and the earth, creation from day three through day six. Following the chronology of creation from Genesis 2:4 through the end of chapter two: the sequencing of events is essentially, 1) the heavens and the earth were created, 2) Adam was created from the dust of the ground, 3) the Garden of Eden was planted and man was place in it.

Where do we go from the time of the planting of the garden? The next event which comes to our attention when reading the chapter is that God brought the animals before Adam to see what he would name them. The naming of the animals, like the herb of the field is a stumbling block for traditionalists.

Theologians nor scholars correctly define the time in creation when the animals <u>are named</u>. They view that this passage (Genesis 2:19-20) relates only and specifically to the naming of the animals and not their creation. Think though, in order for Adam to name the animals would not the animals have to already had been created? When did their creation take place?

Genesis 1:20-25 clearly show when the animals of the earth were created. In verse twenty through twenty-three the creatures of the sea and the birds of the air are are shown to have been created during the fifth day. The animals of the dry land were created during the sixth day.

In the biblical text of Genesis it is stated that the birds, sea creatures and animals were created during the fifth and sixth days of creation. Now, looking back to when Adam named the animals in Genesis 2:19 it can be understood that the naming of the animals would have had to taken place on or after the sixth day. But, wait a minute, the naming of the animals is contextually and chronologically relevant to Genesis 2:18.

The key element to understand the timing of the creation of man is contained in Genesis 2:18. Genesis 2:18 says, *"And the LORD God said, it is not good that the man should be alone; I will make an help meet for him."* Two important items exist in this verse, 1) the man was alone, 2) since the man was alone God would make a helper fit for him. Verse nineteen of chapter two then says, *"And out of the ground the LORD God formed..."*

It is only after God acknowledges that man is <u>alone</u> that he creates the animals of the earth. And since the animals were created in the fifth day and sixth day of creation[56]; and since man already existed, then man had to have been created before the fifth day of creation. Notice that after God creates the animals after seeing that man is alone God sees that, *"...but for Adam there was not found an help meet for him."* (Genesis 2:20). It is at this point that God forms woman from the rib of man.

The chronology of the creation of man therefore is shown as – 1) the man was created (on the third day of creation); 2) the animals of the earth were created (on the fifth and sixth days of creation); 3) woman was created. Woman was created after the creation of the animals on the sixth day of creation and God rested on the seventh day of creation. Looking at this chronology and extend it to its logical conclusion it is seen that woman could only have been created on the sixth day. Is the identification of the creation of man only complete with man's joinder with woman as indicated in Genesis 1:27, *"So God created man in his own image, in the image of God created he him; male and female created he them."*? Well, possibly so. But when looking more closely at the specific details contained in the creation of man on the sixth day, as described in Genesis 1:26-31, it can be found that that this passage actually does not solely relate to the creation of man (Adam) and woman (Eve).

56 Genesis 1:22-25

When man (Adam) was formed from the dust of the ground on the third day as recorded in chapter two of Genesis, why is man created on the sixth day according to chapter one of Genesis? Following the context of creation these descriptions are of two separate creations of two separate entities. A key difficulty to comprehending this fact is that the same noun "adam" is used for both the man formed from the dust (Adam) [Genesis 2:7] and created man [Genesis 1:26]. An additional difficulty to correctly understanding the dual creation of man is contained in Genesis, chapter five.

Verse one of Genesis, chapter five begins, *"This is the book of the generations of Adam..."* This "Adam" is the proper name attributed to the man made from the dust of the ground on day three of creation as described in Genesis chapter two. Yet,, Genesis chapter five continues, *"...In the day that God created man, in the likeness of God made he him; (v.1) Male and female created he them; and blessed them, and called their name Adam, in the day when they were created. (v. 2)."* Genesis five then continues with the lineage of Adam (proper name). Although Genesis chapter five appears to parallel Genesis chapter one verses twenty-six and twenty-seven; "Adam" (proper name) in Genesis chapter five is distinctly different than "adam" (common noun) of Genesis chapter one. Additionally note that if there was only one singular entity "adam" there would be no need to define the lineage of "Adam" (proper name).

There is also two other indicators showing a plurality of man in creation. The first is comparing the restrictions placed on man. The second is understanding the grammatically significance of created "in the image".

Regarding the first indicator, the man created on the third

day (Adam) was formed to till the ground [57] in the garden that God planted in Eden. [58] Adam was limited to location and purpose. In this setting Adam was given authority to eat of all of the trees of the garden EXCEPT the tree of the knowledge of good and evil. [59] In comparison, the man created on the sixth day (adam) was given authority to have dominion over all the fish of the sea, the fowl of the air, the cattle, the earth and everything that creepeth upon the earth [60] God placed no restriction of location on the man created on the sixth day (adam). Furthermore, God gave adam the authority to eat of *"...every tree, in which is the fruit of. a tree yielding seed..."* (Genesis 1:29). Unlike the restrictions set upon Adam; there are no restrictions set upon adam as to location or what could be eaten. These variances may seem inconsequential but, in the biblical narrative of details this is quite important.

Regarding the second indicator, in Genesis chapter one the two verses of focus are verses 26 and 27. In verse 26, "And God said, Let us make man in our image, after our likeness: and let them..." These verses are clear that man was made in the image of God, Yet, also as emphasized earlier, "them" in Genesis 1:26 is stated prior to any introduction of the concept of male and female. This introduction of "them" in verse 26 actually sets the correct understanding of verse 27. Verse 27 reads, *"So God created man in his own image, in the image of God created he him: male and female created he them."* Normal interpretation of this verse is that it merely states the "image" of man in two different ways – in the image of God. However, by using a very valuable grammatical tool of sentence diagramming the true understanding of this verse is concretely tied to the plurality of

57 Genesis 2:5
58 Genesis 2:8
59 Genesis 2: 16-17
60 Genesis 1: 28

"them" in verse 26 and also "them" in verse 27. When diagrammed, the first part of verse 27, *"So God created man in his own image..."* shows that "image" is related to the image of man. Like the animals were created in their "kind", man was created in "his" kind. However, when the second part of of verse 27, *"...in the image of God created he him:"* is diagrammed, man is shown to be created in the image of God. Thus there is a plurality of man in creation – them. Man in the image of man and man in the image of God. The traditional interpretation that these verses show man was conferred with both the image of God and man's own image fails to recognize the identification and import of "them".

Traditionalists hold that mankind was started by a single man and a single woman (Adam and Eve). Yet, when considering the animals is it viewed that each type and species of animal started by a single male and a single female of that type and species? Generally, no. So why is man viewed differently?

The story of Noah provides an underpinning to invalidate the traditionalists point of view as to one singular couple, Adam and Eve, populating the earth from creation to the flood. The summary surrounding Noah was that men of the earth had become wicked and God vowed to destroy man that he had created on the earth via a great flood, *"But, Noah found grace in the eyes of the Lord."*[61] However, when the flood came it was not just Noah that was saved from the flood but also Noah's wife, his three sons and the sons three wives: even though it was only Noah that found grace in the eyes of the Lord. Eight people in total. So, it wasn't exclusively one man and one woman (Noah and his wife) used to repopulate after the flood. Likewise, there was more than one male and female of each species of animal on

---

61 Genesis 6:8

the ark. So, again, did original creation during the six days of creation rely on a single man and a single woman to populate the earth?

Understanding the context of the creation of man (Adam) and (adam) also allows one to properly interpret and understand a couple of difficult biblical passages and concepts. The passages and concepts are: the wife of Cain, and the sons of God and daughters of men referenced in Genesis chapter six.

Genesis chapter four relates the story of two sons of Adam, Abel and Cain wherein Cain kills Abel and is outcast from the presence of the Lord (verse 16) and verse 17 states in part, *"And Cain knew his wife; and she conceived, and bare Enoch:..."* Upon reading this passage, the question arises: from where did Cain's wife originate? With an interpretation of the single man and woman in creation, Adam and Eve – the wife would have had to been a sister of Cain. As such, Cain and his wife would have engaged in one form of action that is known as incest – siblings having sexual intercourse. Most theologians do not view the issue of incest in the context of creation problematic. In adopting this viewpoint it is opined that the negative concept of incest wasn't introduced until the giving of the law to Moses and referenced in Leviticus chapter 18, *"None of you shall approach to any that is near of kin to him, to uncover their nakedness:... (v. 7)."* The particular persons are then described in verses 8 through 16 of Leviticus chapter 18. However, incest is inherently immoral. So, why was it allowed at the beginning of creation? The traditional answer is, "because there was no one else for Cain other than his sister". Incest is inherently immoral even in the beginning of creation. Acceptance of the position that Cain's wife was his sister impugns the character of God. An immutable God does not need to create a variance in morality in order for mankind to propagate. The wife of Cain is fully explainable with the understanding of the plurality of man in creation – Adam made from the dust of the ground on the third day and man on the sixth day.

In Genesis chapter six it is stated, *"And it came to pass, when men began to multiply on the face of the earth, and daughters were born unto them, (v. 1) That the sons of God saw the daughters of men that they were fair; and they took them wives of all which they chose. (v. 2)"* These "sons of God" are also referenced in chapter one and chapter two of the book of Job. So, what or who are these sons of God?

Christian traditionalists opine and teach that these referenced sons of God of the books of Genesis and Job were heavenly creatures, i.e. 'angels'; and by further reference, 'fallen angels'. Yet, Hebrew scholars attribute the sons of God to the lineage of Seth (Adam's third son) being the righteous ones, and inherently more godly in contrast to the sons of Cain (Adam's son and brother of Abel) with the female offspring referenced as the daughters of men.

Although Seth is iterated in the lineage of Adam detailed in chapter five of the book of Genesis; there is still a stark differentiating between the 'sons of God' and the 'daughters of men'. As to the idea of the sons of God referring to angles; the book of Hebrews answers that question in the negative, *"For unto which of the angels said he at any time, Thou art my Son, this day have I begotten thee?"*[62] Although this verse is specifically related to Jesus, the Messiah; the implication is that angels are not referred to as sons of God. However, the answer to who are the sons of God is found in the book of Luke, chapter three. Luke, chapter three starting in verse 24 details the lineage of Jesus of Nazareth, the Messiah. This detailed lineage is

---

62 Hebrews 1:5

69

through Jesus' mother, Mary. Unlike a lineage given in the book of Matthew, this narrative starts with Jesus and works it way backwards identifying an individual's name and clarifying the lineage of the named individual by stating, "which was the son of...". The ending of this lineage is detailed *"...which was the son of Seth, which was the son of Adam, which was the son of God. "*[63]

So, Adam (proper name) made from the dust of the ground on the third day of creation is identified as 'the son of God'. In contrast, the offspring of man created on the sixth day are the 'daughters of men'.

Having addressed two difficult passages and concepts in Genesis (Cain's wife and 'the sons of God' in Genesis chapter six), it is now important to address the theological component of "sin" and Adam in context to the plurality of man in creation. Not surprisingly though, there is no incongruity in scripture.

Although sin is complex in meaning and action (or inaction), and broad in application; the essential definition of sin is missing the mark of God's intended purpose. The origination of sin on earth was Adam's disobedience to God's command and is recorded Genesis chapter three. God's command was specific and is recorded in Genesis chapter two. God's command was given to Adam upon his creation on the third day of creation:

"...And the LORD God took the man, and put him into the garden of Eden to dress it and keep it. (v. 15) And the LORD God commanded the man, saying, Of every tree of the garden thou mayest freely eat: (v. 16) But of the tree of the knowledge of good and evil, thou shalt not eat of it: (v. 17)... " [64]

---

63 Luke 3:38
64 Genesis 2: 15-17

Thus the specific command was given by God to Adam to not of the tree of the knowledge of good and evil. Adam subsequently disobeyed this specific command,

> *"And the woman said unto the serpent, We may eat of the fruit of the trees of the garden: (v. 2) But of the fruit of the tree which is in the midst of the garden, God hath said, Ye shall not eat of it, neither shall ye touch it, lest ye die. (v. 3) And the serpent said unto the woman, Ye shall not surely die: (v. 4) For God doth know that in the day ye eat thereof, then your eyes shall be opened, and ye shall be as gods, knowing good and evil. (v. 5). And when the woman saw that the tree was good for food, and that it was pleasant to the eyes, and a tree to be desired to make one wise, she took of the fruit thereof, and did eat, and gave also unto her husband with her; and he did eat. (v. 6)"* [65]

God gave the specific command to Adam to not eat of the tree of knowledge of good and evil. Adam disobeyed God's specific command to not eat of the tree of knowledge of good and evil. Adam's disobedience to God's specific command resulted in the first occurrence of sin on the earth. This occurrence of sin is affirmed in the book of Romans in the New Testament of the Bible; verse 12 of chapter five states, *"Wherefore, as by one man sin entered into the world..."*

This 'one man' was Adam the man formed from the dust of the ground on the third day of creation. Thus, the plurality of man in creation introduces no contradiction to the theology of the introduction of sin in the earth.

When considering lineages, the lineage of Adam the man formed from the dust of the ground on the third day of creation,

---

65 Genesis 3: 2-6

there are many routes of descendancy available: through Cain, through Seth, or through any one of the many others sons of Adam. And, when considering lineages through man created on the sixth day of creation there are undoubtedly also many variances possible. Bloodlines from the sons of God and the daughters of men were blended for a time. Yet, whether through Adam or through man their lineages culminated at and through Noah and his family. After the great flood recorded in Genesis chapters six through eight, all future bloodlines flowed from Noah, his wife, Noah's three sons, Shem, Ham, Japheth and each of their wives. The bloodlines of mankind from the previously corrupted world continued through eight people. [66]

A logical question may arise, "Since mankind ultimately descended from the eight: Noah, Noah's wife, Noah's son Shem, Shem's wife, Noah's son Ham, Ham's wife, Noah's son Japheth, Japeth's wife; why is it important to understand the plurality of man?" This question is answered from two basic aspects.

One aspect is that the plurality of man provides a complete explanation for the genetic diversity of man evidenced in history and in the world today. The other aspect, primary and most important is that scripture, God's word declares the truth of the plurality of man.

---

66 1 Peter 3:20

CHAPTER 8
O.UTER-SPACE

A common term for all physical items outside the confines of the earth is "space". Although sometimes referred to as "outer-space", the shortened version of "space" is adapted almost universally in speech. Yet, in the totality of space, earth is present within in it. Another way to refer to the totality of space is by use of the words "universe" or "cosmos". However, initially, space is the most generic word to use. Biblically, space is initially described in the book of Genesis, Chapter One, verses one and two, and verses six through 10, and verses 14 through 19:

> *"In the beginning God created the heaven and the earth. (v. 1)And the earth was without form, and void; and darkness was upon the face of the deep. And the Spirit of God moved upon the face of the waters. (v. 2)"*...

> *And God said, Let there be a firmament in the midst of the waters, and let it divide the waters from thew waters. (v. 6) And God made the firmament, and divided the waters which were under the firmament from the waters which were above the firmament: and it was so. (. 7) And God called the firmament Heaven. And the evening and the morning were the second day. (v. 8) And God said, Let the waters under the heaven be gathered together unto one place, and let the dry land appear: and it was so. (v. 9) And God called the dry land Earth; and the gathering together of the waters called he Seas: and God saw that it was good. (v. 10)...*

And God said, Let there be lights in the firmament of the heaven to divide the day from the night; and let them be for signs, and for seasons, and for days, and years: (v. 14) And let them be for lights in the firmament of the heaven to give light upon the earth: and it was so. (v. 15) And God made two great lights; the greater light to rule the day, and the lesser to rule the night: he made the stars also. (v. 16) And God set them in the firmament of the heaven to give light upon the earth, (v. 17) And to rule over the day and over the night, and to divide the light from the darkness: and God saw that it was good. (v. 18) And the evening and the morning were the fourth day. (v. 19)"

The complexity, and simplicity of creation of space is held in these verses. First note that the process took place over the course of three days. The second day of creation through the forth day of creation. As to the simplicity, there are two terms used as the basis for creation: the waters and the firmament. As to complexity, everything created within space and the earth itself is contained in the simplicity of the waters and the firmament.

It is important to recognize that the waters were a preexisting 'element' of creation. Although the earth was initially not yet created, the waters were. In Genesis, Chapter 1, verse 2, it states:

"And the earth was without form and void...And the Spirit of God moved upon the faces of the waters."

From a theological perspective, the waters of creation as a preexisting 'element' of creation does not invalidate God's creation in everything. In the New Testament book of Colossians, in reference to Christ (a person of the triune God), it is stated,

> *"For by him were all things created, that are in heaven, and that are in earth, visible and invisible, whether they be thrones, or dominions, or principalities, or powers: all things were created by him, and for him: (v. 16) And he is before all things and by him all things consist. (v. 17)"*

Within scripture, "all" means "all". So, 'the waters' as a preexisting 'element' of creation of our universe fall within God's creative authority.

Returning to the text of Genesis Chapter 1, verses six through eight,

> *"And God said, Let there be a firmament in the midst of the waters, and let it divide the waters from the waters. (v. 6) And God made the firmament, and divided the waters which were under the firmament from the waters which were above the firmament: and it was so. (v. 7) And God called the firmament Heaven. And the evening and the morning were the second day. (v. 8)"*

In these verses note that the waters' are the entirety of the underlying component from which the physical universe was created. Then 'the waters' were then divided by 'the firmament'. So, after creation of the firmament there were 'the waters' <u>above</u> the firmament and were 'the waters' <u>under </u> the firmament. From this point forward in scripture there is no additional explanation as to what happened to 'the waters' above the firmament. But, there is detailed explanation regarding the firmament and as the waters that were divided under the firmament.

75

This detailed explanation of the waters divided under the firmament are addressed in verses nine and ten of Chapter 1 of the book of Genesis:

> *"And God said, Let the waters under the heaven be gathered together unto one place, and let the dry land appear: and it was so. (v. 9) And God called the dry land Earth; and the gathering together of the waters called he Seas: and God saw that it was good. (v. 10)"*

Please note that in verse nine the word "heaven" is used in the place of the firmament as God had already identified the firmament as "Heaven" is verse eight.

In summary, the narrative of creation of space is 'the waters' above 'the firmament'; the 'firmament', and 'the waters' below 'the firmament'. The waters above the firmament are not further elucidated, the waters under the firmament became the earth – both the seas and the dry land. But, what about 'the firmament'.

It was noted above that in verse eight of Genesis 1 that the firmament was called heaven. In verses 14 through 19 of Genesis 1 additional detail regarding heaven is provided:

> *"And God said, Let there be lights in the firmament of the heaven to divide the day from the night; and let them be for signs, and for seasons, and for day, and years: (v. 14) And let them be for lights in the firmament of the heaven to give light upon the earth: and it was so. (v. 15) And God made two great lights; the greater light to rule the day, and the lesser light to rule the night: he made the stars also. (v. 16) And God set them in the firmament of heaven to give light upon the earth, (v. 17) And to rule over the day and over the night, and to divide the light*

from the darkness: and God saw that it was good. (v. 18)
And the evening and the morning were the fourth day."

The earth exists in heaven -the firmament – and the sun, moon and stars were created and placed in heaven.

By observation, theory and experimentation, the firmament consists of the entirety of the physical universe – both visible and invisible. Yet, what does this entail?

According to current scientific understanding the physical universe includes all the visible matter, antimatter, visible energy, theoretical dark matter, theoretical dark energy, and all the underlying mechanisms and laws for the operation of these components and the universe itself. So far these mechanisms have been classified as and 'fields' and 'forces' and 'dimensions'. Fields are defined as a physical property that can be quantified by measurement at each point in space and time. Forces are related to mass and acceleration and defined as interactions between subatomic particles to change the momentum of an existing particle, to destroy an existing particle, or to create a new particle.[67] Dimensions are the degrees of freedom to move within space.

Dependent upon what endeavor of study, these mechanisms may vary. Classic physics identifies four fundamental forces: electromagnetic force, gravitational force, strong nuclear force and weak nuclear force. Quantum Field Theory, a perception of the interaction of subatomic particles, under its standard model identifies 17 fields being subdivided as 12 matter field, four force fields and the Higgs field.[68] However, depending on the

---

67 https://www.thefreedictionary.com/fundamental+force
68 https://www.quora.com/How-many-quantum-fields-are-there-What-are-they-Does-each-of-them-interact-with-every-other-field?

Thomas W. Guthrie

understanding of what constitutes a field, the number of fields
can range between the standard model 17 fields to 37 fields and
considering degrees of freedom as many as 268 fields.[69] These
fields have names of uniqueness such as quarks, leptons and
bosons. According the study under the String Field Theory, a
perception that the matter and energy are strings and not
particles, there are not fields but dimensions. The standard
model of String Theory identifies 10 dimensions. For a clearer
understanding of the difference between Quantum Field Theory
and String Theory, theoretical physicist Warren Siegel,
Phd,*"The fundamental difference* between *a particle and a
string is that a particle is a 0- dimensional object in space, with
a 1-dimensional world-line describing its trajectory in
spacetime, while a string is a (finite, open or closed) 1-
dimensional object in space, which sweeps out a 2-dimensional
world-sheet as it propagates through spacetime:"[70]* Dr. Siegel's
visual representation of this difference is shown on the following
page.

69 https://physics.stackexchange.com/questions/176941/how-many-quantum-
field-are-there
70 Introduction to String Field Theory, Warren Siegel, University of
Maryland, http://insti.physics.sunysb.edu/~siegel/sft.pdf, page 1, 2

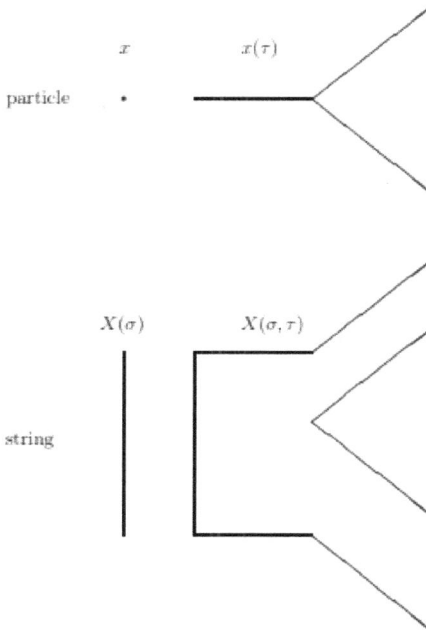

particle x $x(\tau)$

string $X(\sigma)$ $X(\sigma, \tau)$

    Space including all objects in the universe consists of all of these various fields, forces or dimensions, and all the laws of operation that underlie these theoretical particles or strings. Space in earlier times was referred to as the aether, all the material that fills the universe. Biblically, this is the firmament of Genesis.

Thomas W. Guthrie

CHAPTER 9
N.EW CREATION

Within the present creation, current scientific speculation is that all that has been created will cease to exist. In contrast to the Big Bang Theory wherein the universe came into existence via intense heat and energy at a single point in time – a 'singularity'; the universe is theorized to cease existing due to a condition referred to as "heat death", the Big Chill or Big Freeze. This theory is based on the concept of entropy. Entropy is a process understood in physics under the second law of thermodynamics where the randomness of a system increases so that there no longer exists thermal energy to convert to mechanical work. According to the second law of thermodynamics entropy increases with time. Thus, over time there will be no thermal energy in the entirety of the universe.

Of course, there are alternate and opposing scientific viewpoints. For example, "Max Plank wrote that the phrase 'entropy of the universe' has no meaning because it admits of no accurate definition [25][26]" 71 Additionally, "In Peter Landsberg's opinion: 'The *third* misconception is that thermodynamics, and in particular, the concept of entropy can without further enquiry be applied to the whole universe...These questions have a certain fascination, but the answers are speculations.'[32]" 72

None the less, history and observation has shown that the universe does change. There have observations of super

---

71 Https://en.wikipedia.org/wiki/Heat_death_of_the_universe #Opposing_views
72 Ibid

brightening of stars which then disappear. These observations have been defined as supernova, or the exploding of stars. Such observation were historically recorded in ancient China, Japan, Korea and among Arabic astronomers. More recently there have been observations historically recorded such as the supernova of 1006 AD, the supernova of 1054 AD that produced the Crab Nebula [73], Tycho's supernova of 1572 AD, Kepler's supernova of 1604 AD. Observations of debris in the Milky Way Galaxy indicate that a supernova more took place recently, about 140 years ago. [74] It is estimated that there are about two supernova visible from earth every century.

In addition to observed supernova, the number of comets, asteroids and meteorites traveling across the path of the earth have been and are observable. And the evidence of those impacting the earth is observable. These celestial bodies provide additional evidence to the changing universe. Change of this nature should not be a surprise as the earth moves around the sun, the sun moves within the Milky Way Galaxy and the Milky Way Galaxy moves within the universe.

From a biblical perspective earth and the heavens will cease to exist but, will be recreated. Verses ten and 11 of the third chapter of the book of 2 Peter show the manner in which they are destroyed. They state, *"...the heavens shall pass away with a great noise, and the elements shall melt with fervent heat, the earth also and the works that are therein shall be burned up. (v. 10)...Seeing then that all these things shall be dissolved...(v. 11)"* Chapter 65 of the book of Isaiah and chapter 21 of the book of Revelation affirm a new creation. In verse 17 of chapter 65 of the book of Isaiah, the prophet Isaiah writes, *"For behold, I*

---

73 https://academic.oup.com/book/8331
74 https://www.nasa/gov/universe/the-remarkable-remains-of-a-recent-supernova

create new heavens and a new earth..." The book of Revelation, verse one of chapter 21 the apostle John writes, *"And I saw a new heaven and a new earth: for the first heaven and the first earth were passed away;..."*

It should be noted: although there is some relatively new (in the last 50 years) theological interpretation that the passages referenced above define a recreation of the first created, currently existing, heavens and earth; the context and language of the passages fully indicate that he old has been replaced by the new. For example, in the passage quoted from the book of Revelation it is clearly stated that the 'first' heavens and the 'first' earth "...were passed away." Holding to the truth that the first created, currently existing heavens and earth will pass away, Jesus states, *"Heaven and earth shall pass away..."* [75] Additionally: in the book of Isaiah, chapter 51, verse 6, the prophet Isaiah states, *"Lift up your eyes to the heavens, and look upon the earth beneath: for the heavens shall vanish away like smoke, and the earth shall wax old like a garment..."* ; it is shown in the book of Psalm, chapter 102, verses 25 and 26, *"Of old has thou laid the foundation of the earth: and the heavens are the work of thy hands. (v. 25) They shall perish...(v. 26)"*; the book of Hebrews, chapter 1, verses 10 and 11 affirm Psalm 102, verses 25 and 26, *" And, Thou, Lord, in the beginning hast laid the foundation of the earth; and the heavens are the works of thine hand; (v. 25) They shall perish...(v. 26)"*

The remnants of the destruction of planets, suns and even galaxies have been observed via astronomical instruments. And, destruction and death seem to be the observed natural outcome of creation. Why should the outcome of the earth be any different?

---

75 Matthew 24:35, Note also: Mark 31:30 and Luke 21:33

Thomas W. Guthrie

As a precursor to the recreation of the earth, the question of why there is an indicated necessity for recreation of the earth needs to be addressed. From the breadth of the Bible, man is identified as a being created by God for the purpose of fellowship with God. As a being created of flesh, man required an environment and dwelling place suitable for his survival. This place is earth. While it is frequently taught in Christian theology that earth is only a temporary home of man until his death; biblical reality is that the earth is intended to be the permanent home of man. And, for religions that espouse reincarnation, such reincarnations are confined to earth.

From a biblical perspective, the process of man's existence is summarized as, 1) birth, 2) life [encompassing all of life's decisions], 3) death, 4) resurrection from the dead [to life with God or life apart from God], 5) eternal existence. It is during step 2) above that the significance of earth and the recreated earth is realized.

The book of Revelation, chapter 21 details the eternal existence. The underlying emphasis of the recreated earth is shown in verses one through three, *"And I saw a new heaven and a new earth: for the first heaven and the first earth were passed away, and there was no more sea. (v. 1)...And I heard a great voice out of heaven saying, Behold, the tabernacle of God is with men, and he will dwell with them, and they shall be his people, and God himself shall be with them, and be their God. (v. 3)"* It needs to be noted that this eternal existence with God on the recreated earth is only applicable to those whose who have their names written in the 'book of life'. This 'book of life' is God's recording of all men that have acknowledged and confessed Jesus of Nazareth as the Son of God and believed and accepted Jesus' death, burial and resurrection from the dead.

84

This acknowledgment, confession, belief and acceptance is commonly referred to as salvation, or being saved. The book of Romans in chapter 10, verses nine and 10 is one of the most forthright scriptures portraying the process and outcome of acknowledgment, confession, belief and acceptance, *"That if thou shalt confess with thy mouth the Lord Jesus, and shalt believed in thine heart that God hath raised him from the dead, thou shalt be saved, (v. 9) For with the heart man believeth unto righteousness; and with the mouth confession is made unto salvation. (v. 10)"*

A broader understanding of 'Lord Jesus' can be found in the 4th chapter of the book of Acts. The apostle Peter upon being brought before the Jewish high priest, Annas and others after healing a man, declares, *"Be it known unto you all, and to all the people of Israel, that by the name of Jesus Christ of Nazareth, whom ye crucified, whom God raised from the dead, even by him doth this man stand here whole. (v. 10) This is the stone which was set at nought of you builders, which is become the head of the corner. (v. 11) Neither is there salvation in any other: for there is none other name under heaven given among men, whereby we must be saved. (v. 12)"*

This acknowledgment, confession, belief and acceptance is one of the decisions of life: and the only decision that determines the place of your eternal existence. Will you reside with God on a recreated earth?

Thomas W. Guthrie

CHAPTER 10
S.UBLIME

Within creation detailed in the Bible there are components of creation that are sublime. While sublime they are also often subtle. In the sublime and subtlety these components are either discounted by science or overlooked. Yet, whether observed or not, studied or not, discounted or not; these components are a part of creation.

SOVEREIGNTY

Sovereignty is the complete power and authority which is imputed to an entity. In creation this entity is God as he defines himself in scripture. God is the sovereign in that he imputes to himself the complete power and authority of creation, its existence, its future and its operation. And due to his sovereignty he determined the order of creation and which details of creation he has chosen to reveal to man.

Most of the recognized details of creation revealed to man occur in the book of Genesis. These details include the time of creation – 7 literal 24 hour days; the order of creation – light, earth, foliage on the earth, sun, moon, stars, sea creatures, and beasts of the earth,

God in the book of Job has a discourse with Job in a seemingly rhetorical manner questioning Job's understanding of God's sovereignty in creation. This discourse challenges Job's understanding of God and

God's creation and it should challenge the readers also. It should also challenge scientists because it is beyond merely philosophical or theological as it provides clues to understanding creation. A portion of this discourse in chapter 38 of the book of Job reads, in part:

"Where wast thou when I laid the foundations of the earth (v. 4)...Who hath laid the measures thereof...or hath stretched the line upon it?(v. 5) Whereupon are the foundations thereof fastened? Or who laid the corner stone thereof; (v. 6)...Or who shut up the sea with doors, when it brake forth... (v. 8)...Have the gates of death been opened unto thee? Or hast thou seen the doors of the shadow of death? (v. 17)...Has thou perceived the breadth of the earth?...(v. 18)...Where is the way where light dwelleth? And as for darkness, where it the place thereof, (v. 19)...By what way is the light parted, which scattereth the east wind upon the earth? (V. 24)"

Could the question of the fastening of the foundations of the earth be a reference to classic gravity and orbital mechanics? Or, might it allude to string theory? Or, it there another mechanism such as specific quantum fields not yet explored? Whether likely or not: God is sovereign in his creation. Creation is sublime.

SERPENT

The serpent in chapter 3 of the book of Genesis is an enigmatic entity. Created as a beast of the field [76] this beast of the field communicated with the woman [77] which God had created. And, the woman communicated with

---

76 Genesis 3:1
77 Genesis 3:1,4 and 5

the serpent. [78] At the time of joint communication the serpent was not in a form which we currently recognize as being a serpent, a snake. With this unrecognizable form it is understandable that current man also cannot recognize that the serpent could and did communicate with the woman.

One approach to understanding the serpent of Genesis 3 is to present the serpent as an allegory to describe the temptation the woman felt to eat the fruit of the tree of the knowledge of good and evil. But, allegorical interpretation is an affront to scripture.

Another approach in an attempt to understand the serpent of Genesis, chapter 3 is to ascribe a different meaning to the Hebrew word for serpent. Such change is to define the word for serpent as 'shining' and is related to its usage in the book of Numbers, chapter 21 wherein Moses was instructed by God to make a serpent of brass and put it on a pole.[79] Accordingly, the shining of brass is inextricably claimed to be intertwined with serpent. Thus, an interpretation of the serpent being a shining angelic being is presented, or as the apostle Paul states, *"For such are false apostles, deceitful workers, transforming themselves into the apostles of Christ. (v.13) And no marvel; for Satan himself is transformed into an angel of light. (v. 14) "*[80] However, in the book of Genesis the serpent is clearly identified as a beast of the field. And a beast of the field is not an angelic being and an angelic being is not a beast of the field. Thus, an angelic being interpretation of the serpent is not

---

78 Genesis 3:2 and 3
79 Number 21:8,9
80 2 Corinthians 11:13,14

justifiable.

However, in the Bible there is another account of a beast of the field communicating with man. In chapter 22 of the book of Numbers, verses 28 through 30 there is an exchange between a prophet named Balaam and his ass which he was riding: *"And the LORD opened the mouth of the ass, and she said unto Balaam, What have I done unto thee, that thou hast smitten me these three times? (v.28) And Balaam said unto the ass, Because thou has mocked me: I would there were a sword in mine hand, for now would I kill thee. (v. 29) And the ass said unto Balaam, Am not I thine ass, upon which thou hast ridden ever since I was thine unto this day? was I ever wont to do so unto thee? And he say, Nay. (v. 30)"* This account in Numbers shows that the LORD opened the mouth of the ass. But, from the context of the exchange it was the ass communicating with Balaam and not God communicating through the ass. And, while this seems odd or impossible it is unknown as to the full relationship between man and beasts of the field especially at the front-end of creation. Such communication between the woman and the serpent should therefore not be surprising.

The text in Genesis chapter three also states that the serpent was more subtil than any other beast of the field. [81] Subtil has the meaning of being crafty, sly or shrewd and perhaps even deceptive. This entity therefore had a propensity for or characteristics of being crafty. This craftiness and deceptiveness is commonly associated with Satan, or the Devil. While Jewish theology does not associate the serpent in Genesis with Satan or the Devil,

---

81 Genesis 3:1

mainstream Christian theology does. Such association is intimated by Revelation 12:9 *"And the great dragon was cast out, that old serpent called the Devil, and Satan, which deceiveth the whole world..."* and Revelation 20;2, *"And he laid hold on the dragon, that old serpent which is the Devil, and Satan..."* Mainstream Christian theology portrays the serpent as being influenced by Satan. Within this line of thinking, God's judgment against the serpent is two-fold. First God declares, *"And the LORD God said unto the serpent Because thou hast done this, thou art cursed above all cattle, and above every beast of the field; upon thy belly shalt thou go, and dust shalt thou eat all the days of thy life;"*[82] Secondly while still talking to the serpent God states, *"And I will put enmity between thee and the woman, and between thy seed and her seed; it shall bruise thy head, and thou shalt bruise his heel."*[83] The first judgment as detailed in verse 14 of the third chapter of Genesis relates to the physical judgment of the serpent and characteristics of this entity's being that will be changed by this judgment of God – upon thy belly it shalt go and eat the dust all the days of its life. The second judgment as detailed in verse 15 of the third chapter of Genesis relates to the influence by the entity Satan, or the Devil upon the serpent's being but; the judgment is against Satan. While it is common to assess that a serpent generally strikes the heel (or leg) of a human and a human is prone to retaliate or preemptively attack, or bruise, the head of a serpent; the judgment in this verse is a prophetic declaration referring to the coming of the Jewish Messiah, Jesus of Nazareth and upon the Messiah's resurrection the Messiah's overcoming the deception and accusations by Satan

---

82 Genesis 3:14
83 Genesis 3:15

against mankind. Three key characteristics of the prophetic declaration are affirmed in history – 1) 'Seed' is of man not of woman. But here seed is of the woman: this affirms the virgin birth of Jesus of Nazareth to Mary. 2) The 'seed' is referred to 'it'. Thus it is singular in nature and not indicative of lineage. 3) 'Bruising' of 'his' heel is referencing the bruising of Jesus' heel by his crucifixion on the cross. And, 'his' also dictates singularity.

It is clearly defined in scripture that the serpent was: 1) a beast of the field, 2) subtil, and 3) could communicate with the woman. Also, God in his sovereignty changed the physical attributes of his created being, the serpent. Accordingly, the proper approach to understanding the serpent of Genesis 3 is to accept scripture as written. Even though modern man believes he can (and does) know all of creation, such belief is not so and creation is sublime.

SIN

Sin is a difficult concept to grasp. Sin is a difficult reality to grasp.

Sin is sometimes referred to in simple conceptualization as "missing the mark" as in an archer missing the bullsyey on a target. But, what is 'the mark'? How is 'the mark' determined? Who determines 'the mark'? How does one know if they have "missed the mark"? Although mankind has collectively and individually an innate feeling of sin and its presence or absence, it is through religious beliefs wherein one begins to hone their understanding of sin.

In Christianity the Bible also referenced as scripture provides the groundwork for the understanding the concept and reality of sin. In verse 12 of chapter five of the book of Romans it states in part, *"Wherefore, as by one man sin entered into the world..."* This verse is in reference to the disobedience of the man commonly referred to as Adam. Adam was given a directive by God to not eat of the fruit of the tree of the knowledge of good and evil.[84] Adam disobeyed God's directive. [85] Thus, by this action of disobedience sin entered into the world.

There are many examples of sin in the Bible. However, a honing of what sin is can be distilled from four scripture verses: James 1:14-15 states, *"But every man is tempted, when he is drawn away of his own lust, and enticed. (v. 14) Then when lust hath conceived, it bringeth forth sin:... (v. 15)"*, James 4:17 shows, *"Therefore to him that knoweth to do good, and doeth it not, to him it is sin."*, It is written in 1 John 3:4 *"...for sin is the transgression of the law."*, in 1 John 5:17 it is stated, *"All unrighteousness is sin."*

Adam in his decisions and actions exemplified the meaning of sin as elaborated in the above four scripture citations. It is shown that at the time of Adam sinning there was only one law – to not eat of the tree of the knowledge of good and evil – for sin is the transgression of the law (1 John 3:4). Inherently and intuitively knowing that doing good was paramount as doing good was obedience to God's command – to not eat of the tree of the knowledge of good and evil - by not doing it, it became sin unto him (James 4:17). In verse six of

---

84 Genesis 2:9
85 Genesis 3:17

Genesis chapter 3 lust enters into the situation, *"And when the woman saw that the tree was good for food, and that it was pleasant to the eyes, and a tree to be desired to make one wise, she took the the fruit thereof, and did eat, and gave also unto her husband with her, and he did eat."* Lust overtook the man and the woman – to not eat of the tree of the knowledge of good and evil - But every man is tempted, when he is drawn away of his own lust, and enticed. Then when lust hath conceived, it bringeth forth sin. (James 1:14-15). Righteousness is commonly conceived as being in the 'right standing' with God. Therefore, unrighteousness is not being in the right standing with God. Unrighteousness was exhibited upon disobeying the command - to not eat of the tree of the knowledge of good and evil - all unrighteousness is sin. (1 John 5:17)

Adam's decisions and actions were individually personal to him and resulting in consequences for him. The serpent made decisions and actions resulting in consequences for the serpent. Likewise, Satan's decisions and actions resulted in consequences for Satan And, likewise the woman's decisions and actions resulted in consequences for the woman. Therefore, with multiple entities involved, why is it stated that sin entered into the world by one man? It was Adam's action of eating the fruit of the tree of knowledge of good and evil whereby sin entered into the world by one man. The disobedience to eat and the eating both impacted creation. Not only were there God decreed consequences for each entity involved (man, woman, serpent, and Satan), animals were killed by God as a covering in order to clothe Adam and his wife.[86] However, this killing of animals also

---

86 Genesis 3:21

functioned not only as physical covering for Adam and his wife; it served as a covering for the committed sin. In the book of Hebrews, chapter 9, verse 22 it states in part, *"...without shedding of blood there is no remission."* In context this verse in the book of Hebrews is looking back at the sacrifices dictated in the Mosaic law of the Jewish people of the old testament of the Bible. But, the shedding of blood for the remission of sin via animal sacrifices was practiced throughout history by seemingly all cultures and predated the Mosaic law. Man knew blood sacrifice was required for sin, because God provided an example to Adam upon Adam's initial and original sin. Yet, animal sacrifices only provide a temporary remission of sin. With a need for animal sacrifices; man's relationship with animals changed causing a cascading event of corruption within creation.

While Adam's sin is generally referred to or viewed as a 'spiritual' issue; sin extends beyond 'the spiritual' to man's physical being. As known from historical human experience, scientific observation and scientific experimentation; ingestion of certain botanicals can cause mind-altering and/or mind-enhancing effects. Ingestion of certain botanicals can also produce other physiological effects as well. Science has identified that physiological changes can include changes to the very essence of man on his genes and DNA. Upon the man's eating of the tree of the knowledge of good and evil, *"...the LORD God said, Behold, the man is become as one of us, to know good and evil:..."*[87] The man ingested a mind-altering, mind-enhancing botanical and his knowledge and world view was forever changed. This change had so much impact that God expelled the man

---

87 Genesis 3:22

from his location of creation, *"...and now, lest he put forth his hand, and take also the tree of life, and eat, and live forever: (v. 22) Therefore the LORD God sent him forth from the garden of Eden...(v. 23)"*[88] Adam's sin resulted not only in his understanding of good and evil, his eviction from the Garden of Eden but, most importantly: his genetic alteration from eating the fruit from the tree of the knowledge of good and evil was passed to all his lineage throughout history. Thus, the entirety of mankind has a condition that is referred to as a 'sin nature'.

SOUL

Soul sometimes refers to an individual entity and most generally to a human or a beast of the field. At its core though soul is the life essence of a being. An internet inquiry for 'what is the soul' through bing.com generated an AI response, "The soul is an immaterial essence that is often associated with a person's idenity, personality, and memories. It is considered to be the non-material aspect of a living being that can survive physical death. From a philosophical perspective, the soul is the product of consciousness and self-perception, while sprituality views it as a divine creation connecting us to something greater."[89]

A first look at soul is found in the Bible, the book of Genesis, chapter 2, verse 7, *"And the Lord God formed man of the dust of the ground, and breathed into his nostrils the breath of life; and man became a living*

---

88 Genesis 3:22-23
89 Bing.com cited references from wikipedia.org, britannica.com and thewitness.org

soul." While the soul is associated with the body it is a distinct and separate component of an entity. For example, in the book of Matthew, chapter 10, verse 28 it states, *"And fear not them which kill the body, but are not able to kill the soul: but rather fear him which is able to destroy both soul and body in hell."* And the Bible also states, *"...I pray God your whole spirit and soul and body be preserved blameless..."*[90] This passage shows that not only is the soul distinctively different from the body but, it is also distinctively different from the spirit.

The soul is sometimes referred to as a non-material, and/or a non-physical aspect to a created being. However, by its existing; the soul is in fact not non-material nor non-physical. The soul of created beings are real to the created universe and dwell therein.

SPIRIT

Spirit is more difficult to discern than even soul. Generally spirit is defined as a force or principle animating a being. While a soul is specifically 'attached to' and a component of a created being, spirit is a aspect of creation with multiple points of reference. The discussion of spirit in this section will be primarily from the perspective of the Bible as it provides a broader understanding of spirit.

In the Bible the first reference to 'spirit' is in verse two of Genesis chapter one, *"...And the Spirit of God moved upon the face of the waters."* In the New Testament of the Bible, the book of John, Jesus states,

---

90 1 Thessalonians 5:23

"God is a spirit..."[91] So, it is identified that God embodies spirit.

Later in the book of Genesis, chapter six, verse three it is stated, *"And the LORD said, My spirit shall not always strive with man..."* This verse indicates that the Spirit of God interacts with man. This interaction by the Spirit of God can be broadly applicable to man or confined to individuals. And, as with the Spirit of God above referenced in Genesis 6:3, the Spirit of God can be removed. The book of First Samuel in chapter 11 shows both the infusion of the Spirit of God into an individual and the removal of the Spirit of God from an individual: verse 6, *"And the Spirit of God came upon Saul..."*; verse 14, *"But the Spirit of the LORD departed from Saul..."* Throughout the period of the writings defined as the Old Testament; the Spirit of God was infused into kings and prophets and also removed from them. In Psalm 51, verse 11 King David acknowledging this infusion and removal writes, *"...and take not thy holy spirit from me."* The Spirit of God evidently also outwardly manifests himself in ways recognizable by man. In the New Testament of the Bible the books of Matthew, Mark and John each make testament that John the Baptist saw the Spirit of God descending on Jesus like a dove. For example, in verse 16 of chapter three of the book of Matthew it is written, *"And Jesus, when he was baptized, went up straightway out of the water: and, lo, the heavens were opened unto him, and he saw the Spirit of God descending like a dove, and lighting upon him:"* In the book of John, chapter one, verse 32 there is testimony, *"And John bare record, saying, I saw the Spirit descending from heaven like a dove, and it abode upon*

---

91 John 4:24

him:" Two separate individuals, Jesus and John the Baptist witnessed the Spirit of God as being 'like a dove' in a physical manifestation of the Spirit of God.

Apart from the Spirit of God, (also known as the Holy Spirit) is the spirit of man. Although the spirit is a component part of man as is the soul and body; many times the understanding of spirit and soul are convoluted.

The spirit of man is most often perceived as an individual's thoughts and emotions. The clearest distinction in the Bible delineating that the spirit is not associated with a man's thoughts may be in the book of Daniel, chapter seven, verse 15, *"I Daniel was grieved in my spirit in the midst of my body, and the visions of my head troubled me."* In this verse Daniel distinguishes that his thoughts (the visions in his head) were different from the grief in his spirit (in the midst of his body). An affirmation of the spirit of man being a component, or aspect, of man's (humankind's) being is found in 1 Chronicles, chapter five, verse 26, *"And the God of Israel stirred up the spirit of Pul king of Assyria, and the spirit of Tilgathpilneser king of Assyria..."* Pul the king of Assyria and Tilgathpilneser undertook actions against certain of the Israelites each by the own volition of their being, their spirit. An added aspect of the singular, individual association of the spirit of man to a single individually created man is an example found in the book of 1 Samuel, chapter 28. The prophet, Samuel who had been counsel to king Saul had died and was buried.[92] However, after Samuel's death Saul still desired counsel of him because God was not answering Saul's entreaties. Therefore, in verses 7 and 8 of chapter 28 of of 1 Samuel

92 1 Samuel 28: 3

it is written, *"Then said Saul unto his servants, Seek me a woman that hath a familiar spirit, that I may go to her, and enquire of her. And his servants said to him, Behold, there is a woman that hath a familiar spirit at Endor, (v. 7) And Saul disguised himself, and put on other raiment, and he went, and two men with him, and they came to the woman by night: and he said, I pray thee, divine unto me by the familiar spirit, and bring me him up, whom I shall name unto thee. (v. 8)"* The person Saul named was the deceased prophet Samuel wherein Samuel (in physical appearance) appeared to the woman, and Samuel spoke with Saul.[93] Thus the spirit of Samuel who was still recognizable as Samuel and with the knowledge of Samuel was brought up from the grave.

Spirit in several languages is distilled from the idea of 'breath' or 'wind'. In precept the 'breathing' of an organism is evidence of life and the spirit can, like a 'wind' go outside of the organism. In definition though, the spirit is the animating of an physical organism to give it life. However, if this definition is accurate there can be no separation of a man's spirit from his body and soul. And yet, in scripture there are glimpses of the spirit of man being outside the body. For example, *"So he carried me away in the spirit into the wilderness:..."*[94] The apostle Paul in the book of 2 Corinthians, chapter 12, verse two declares, *"I knew a man in Christ above fourteen years ago, (whether in the body, I cannot tell; or whether out of the body, I cannot tell: God knoweth;) such an one caught up to the third heaven."* In this passage it is intimated that the man could have been translated to the third heaven in the body, or in the spirit.

---

93 1 Samuel 28: 11-19
94 Revelation 17:3

With the scriptural background knowledge that the spirit of man can reside outside the body; it should not be surprising that some people attempt to practice separation of the spirit from the body for 'out of body experiences' via methods such as meditation or astral projection.

Beyond the Spirit of God and a man's spirit there exists other 'spirits'. Many are familiar with terms such as, "evil spirit" or "a lying spirit" but what are these 'spirits'?While perceptually these 'spirits' are many times attributed as thoughts or internal manifestations by an individual, or manifestations and actions against an individual by Satan: these 'spirits' are entities created by God.

A concise picture of these created entities is found the the Old Testament book of 2 Chronicles. The contextual background is that a king of Israel, Ahab is attempting to do something contradictory to counsel of the prophet of God, Micaiah. Micaiah affirms his counsel to Ahab by relaying details of his counsel, *"Again, he said, Therefore hear the word of the Lord; I saw the Lord sitting upon his throne, and all the host of heaven standing on his right hand and on his left. (v. 18) And the Lord said, Who shall entice Ahab king of Israel, that he may go up and fall at Ramoth-gilead? And one spake saying after this manner, and another saying after that manner. (v. 19) Then there came out a spirit, and stood before the Lord, and said, I will entice him, Wherewith? (v. 20) And he said, I will go out, and be a lying spirit in the mouth of all his prophets. And the Lord said, Thou shalt entice him, and thou shalt also prevail: go out, and do even so. (v. 21)"*[95] Many of the negative attributes of

---

95 2 Chronicles 18: 18 -21

mankind according to the science of psychology are associated with an individual's psyche: evil, lying, jealousy... Yet, according to scripture referenced above in 2 Chronicles these attributes are affiliated with created entities: *"Or when the spirit of jealousy cometh upon him.."*[96], *"...and I will be a lying spirit..."*[97], *"...and an evil spirit from the Lord troubled him..."*[98], *"And an evil spirit answered and said, Jesus I know, and Paul I know, but who are ye?"*[99] Likewise in the verses 16 through 18 of the 16th chapter of the book of Acts the apostle Paul recognizes a woman possessing a 'spirit of divination' which is subsequently cast out of the woman by him. A particularly intriguing spirit is the concept and reality in scripture of a 'familiar spirit'. Scriptural references show that some living individuals can communicate with the spirit of deceased individuals by way of a 'familiar spirit'[100]. Familiar spirits like the other spirits (e.g. evil, jealousy, lying, divination) are created entities.

Humankind is a conduit or repository for and with other spirits. The Spirit of God can reside within humankind. The Spirit of God can be withdrawn from within humankind. Likewise, the created entities referred to as spirits can reside within humankind and can be removed from humankind. Yet, the fundamental importance of this reality is that the spirit of man can and does interact with the Spirit of God and the created spirits. Scripture affirms this interaction, *"For we wrestle not against flesh and blood, but against principalities,*

96 Numbers 5:25
97 1 Kings 22:22
98 1 Samuel 16:14
99 Acts 19:15
100 1 Samuel 28:8

against powers, against the rulers of darkness in this world, against spiritual wickedness in high places. "[101]

SALVATION

Salvation as a general meaning is, "preservation or deliverance from harm, ruin, loss, or calamity".[102] From a Christian theological perspective it is the act of being saved from the wrath of God due to an individual's unrighteousness. Christian Biblical salvation is a simple act of faith in Jesus of Nazareth, who he is and what he did to redeem an individual from the consequences of sin – both individual sin and sin derived from Adam. Importantly though, this single avenue of salvation is universally applicable to all mankind regardless of existing individual religious belief.

The simplicity of salvation is found in the book of Romans, Chapter 10, verse nine, *"That if thou shalt confess with thy mouth the Lord Jesus, and shalt believe in thine heart that God hath raised him from the dead, thou shalt be saved.* "[103] The necessity of salvation is for the restoration of mankind to a proper relationship with God the creator, and restoration of creation itself from the result of Adam's initial disobedience and eating of the fruit of the tree of the knowledge of good and evil.

Mankind understands this requirement for redemption and salvation in his soul but, in most cases does not understand and accept the singular mandatory

---

101 Ephesians 6:12
102 Oxford English Dictionary –
   https://www.oed.com/dictionary/salvation_n#24436438
103 Romans 10:9

avenue for salvation. Sacrifices to a god or to gods, doing what is perceived as the right or proper action or actions, honor and death in battle, martyrdom, thinking the right thoughts, etc. are all paths perceived to be acceptable to obtain the salvation required. Some even adopt a perspective that salvation can only be acquired by multiple or endless birth and death life-cycles through a process called reincarnation in order to become perfect and thus obtain salvation. However, from a biblical basis salvation has but one mandatory avenue, Jesus. The apostle Peter in speaking of Jesus under the power of the Holy Spirit declares, *"Neither is there salvation in any other: for there is none other name under heaven given among men, whereby we must be saved."*[104]

Of course to understand who is Jesus of Nazareth, Jesus Christ, Lord Jesus: an individual must be willing to accept what the Bible says of him.

Jesus was a historical person. Born in the region of the world now referred to as the Middle East in the area then known as Judaea. He was born in the village Bethlehem,[105] as an infant was present in the city of Jerusalem[106] and was raised in and hailed from the town of Nazareth[107] and is thus referred to as Jesus of Nazareth. He was born in the year now referred to as 4 B.C.[108] Jesus was crucified and died in public in 33 A.D.[109]

---

104 Acts 4:12
105 Luke 2:4-7
106 Luke 2:22-32
107 Luke 2:39
108 Guthrie, T.W. (2003) *Birth Basics, When Jesus was Born*, Thomas W. Guthrie
109 Jesus ministered for 3-1/2 years after John the Baptist's ministry began in

The controversy regarding Jesus of Nazareth is related to his birth and his resurrection from the dead after being crucified. Jesus is a man conceived by the Spirit of God within a virgin woman.[110] Upon conception a child receives 100% of the essence of the father and 100% of the essence of the mother. Accordingly, Jesus is fully divine and fully human. Joseph the husband of Mary contemplated his wife's pregnacy and had the paterity of Jesus validated by an angel in a dream, *"But while he thought on these things, behold, the angel of the Lord appeared unto him in a dream, saying, Joseph, thou son of David, fear not to take unto thee Mary thy wife: for that which is conceived in her is of the Holy Ghost. (v. 20) And she shall bring forth a son, and thouth shalt call his name JESUS: for he shall save his people from their sins. (v. 21) Now all this was done, that it might be fulfilled which was spoken of the Lord by the prophet, saying, (v. 22) Behold, a virgin shall be with child, and shall bring forth a son, and they shall call his name Emmanuel, which being interpreted is, God with us.[111] (v. 23) Then Joseph being raised from sleep did as the angel of the Lord had bidden him, and took unto him his wife: (v. 24) And knew her not till she had brought forth her firstborn son: and he called his name JESUS. (v. 25) "*[112] Jesus throughout scripture is identified as the Son of God. During his ministry in Judea before his crucifixion: devils possessing two individuals recognized Jesus as the Son of God, *"And when he was come to the other side unto the country of the Gergesenes, there met him two*

---

29 A.D.
110 Luke 1:34-35
111 Isaiah 7:14
112 Matthew 1:20-25

possessed with devils, coming out of the tombs, exceeding fierce, so that no man might pass by that way. (v. 28) And, behold, they cried out, saying, What have we to do with thee, Jesus, thou Son of God?... (v. 29) "[113]; his disciples recognized him as the Son of God, *"He saith unto them, But whom say ye that I am? (v. 15) And Simon Peter answered and said, Thou art the Christ, the Son of the living God"*[114]; the Jewish high priest recognized who Jesus was and questioned Jesus for affirmation, *"...And the high priest answered and said unto him, I adjure thee by the living God, that thou tell us whether thou be the Christ, the Son of God. (v. 63) Jesus saith unto him, Thou hast said:...(v. 64) "*[115] It is even attested that God affirmed Jesus is his son, *"Now when all the people were baptized, it came to pass, that Jesus also being baptized, and praying, the heaven opened, (v. 21) And the Holy Ghost descended in a bodily shape like a dove upon him, and a voice came from heaven, which said, Thou art my belived Son; in thee I am well pleased. (v. 22) "*[116]

As to Jesus' resurrection it is written, *"Then the same day at evening, being the first day of the week, when the doors were shut where the disciples were assembled for fear of the Jews, came Jesus and stood in the midst, and saith unto them, Peace be unto you. (v. 19) And when he had so said, he shewed unto them his hands and his side. Then were the disciples glad, when they saw the Lord. (v. 20) "*[117]*..."And after eight days again his disciples were within, and Thomas with them: then came*

113 Matthew 8:28-29
114 Matthew 16:15-16
115 Matthew 26:63-64
116 Luke 3:21-22
117 John 20: 19-20

Jesus, the doors being shut, and stood in the midst, and said, Peace be unto you."[118]*..."Jesus then cometh, and taketh bread, and giveth them, and fish likewise. (v. 13) This is now the third time that Jesus shewed himself to his disciples after that he was risen from the dead. (v. 14)"*[119]*..."* The apostle Paul fortifies the reality of Jesus' resurrection in the book of 1 Corinthians, *"For I delivered unto you first of all that which I also received, how that Christ died for our sins, according to the scriptures; (v. 3) And that he was buried, and that he rose again the third day according to the scriptures: (v. 4) And that he was seen of Cephas, then of the twelve: (v. 5) After that, he was seen of above five hundred brethren at once;... (v. 6)"*[120]

The historic Jesus of Nazareth's birth was acknowledged publicly and openly by wise men from the East[121] and from the shepherds around Bethlehem at his birth.[122] His crucifixion was public and open in Jerusalem and although his physical appearances resurrection were privately revealed to just his followers; his resurrection was openly affirmed by the empty burial sepulchre and the amazement and consternation surrounding it...

"In the end of the sabbath, as it began to dawn toward the first day of the week, came Mary Magdalene and the other Mary to see the sepulchre. (v. 1) And behold, there was a great earthquake: for the angel of the Lord descended from heaven, and came and rolled back the stone from the door, and sat upon it. (v.2) His

---

118 John 20:26
119 John 21: 13-14
120 1 Corinthians 15:3-6
121 Matthew 2:2-3
122 Luke 8-17

Thomas W. Guthrie

countenance was like lightning, and his raiment what as snow: (v. 3) And for fear of him the keepers did shake, and became as dead men. (v.4) And the angel answered and said unto the women, Fear not ye: for I know that ye seek Jesus, which was crucified. (v. 5) He is not here: for he is risen, as he said. Come, see the place where the Lord lay. (v. 6) "[123] In the book of Luke it is written, *"Now upon the first day of the week, very early in the morning, they came unto the sepulchre, bringing the spices which they had prepared, and certain others with them. (v. 1) And they found the stone rolled away from the sepulchre. (v. 2) And they entered in, and found not the body of the Lord Jesus. (v. 3) And it came to pass, as they were much perplexed thereabout, behold, two men stood by them in shining garments: (v. 3)* "[124]

Although the accounts of the empty tomb appear to reflect only the followers of Jesus: details of his placement in the sepulchre involve other people. In the ciation from Matthew 28, verse four refernced above, *"And for fear of him the keepers did shake, and became as dead men"*: these 'keepers' are Roman soldiers stationed at the sepluchre to prevent the disciples from stealing the body of Jesus. These soldiers were stationed at the sepulchre by command of the governor Pontius Pilate at the request of the chief priests and Pharisees of the Jews.[125]

The reason for the advent of Jesus can be summarized from the words of Simeon during Jesus' visit to Jerusalem during his infancy, *"For mine eyes have*

---

123 Matthew 28: 1-6
124 Luke 24: 1-4
125 Matthew 27: 62-66

108

seen thy salvation, (v. 30) Which thou hast prepared before the face of all people; (v. 31) A light to lighten the Gentiles, and the glory of thy people Israel. (v. 31)"[126] Mankind cannot 'save' itself; either individually or collectively.

Jesus; fully man and fully God: is the only avenue of salvation... "...that God was in Christ, reconciling the world unto himself, not imputing their trespasses unto them...(v. 19) .."For he hath made him to be sin for us, who knew no sin; that we might be made the righteousness of God in him. (v.21)"*[127]

SERAPHIM & SUCH

'Host' is a reference throughout the Bible to created entities. The created entity is usually associated with a group of like kind and is generally associated with a large group, such as an army. For example, in the book of 2 Kings, chapter 6, verse 24 it is stated, *"And it came to pass after this, that Benhadad king of Syria gatherd all his host, and went up, and besieged Samaria."* This passage clearly refers to the Syrian king's army. In the book of Genesis, chapter 32, verses 1 and 2, Jacob encounters angels of God and identifies them as "God's host".

The acknowledgment of these entities referenced by Jacob as "God's host" is biblically identified in the first part of the book of Genesis in chapter two. Upon completion of creation it is stated, *"Thus the heavens*

126 Luke 2: 30-32
127 2 Corinthians 5: 19, 21

109

and the earth were finished, and all the host of them. "[128]
This statement specifically provides a phrase containing
three nouns: the heavens, the earth and the host. And the
host is associated with both the heavens and the earth.
However, it should be noted that the host is not
referenced 'stars' as these created items are constituent in
and being a component of the heavens. This delineation
between 'stars' and 'the host of heaven' is shown in the
biblical books of Deuteronomy and 1 Kings. In
Deuteronomy 4:19 Moses writes God's admonishment to
the Israelites, *"And lest thou lift up thine eyes unto
heaven, and when thou seest the sun, and the moon, and
the stars, even all the host of heaven, shouldest be driven
to worship them, and serve them, which the LORD thy
God hath divided unto all nations under the whole
heaven."* The shows a distinction between the stars and
the host of heaven. In 1 Kings 22:19 it is recorded that
the prophet Micaiah told the king of Israel, *"And he said,
Hear thou therefore the word of the Lord: I saw the Lord
sitting on his throne, and all the host of heaven standing
by him on his right hand and on his left."* In this passage
the host of heaven is clearly identified as being created
entities and not stars.

In Genesis chapter two, verse 1 the host related to
the earth is obviously the created beasts, fish, fowls of
the air, creeping things, and man. The host of the heavens
is however, nuanced. Yet some illumination is given in
scripture.

In creation, 'heaven' is distinguishable from 'the
heavens'. On the second day of creation, *"God called the*

---

128 Genesis 2:1

firmament Heaven. "[129] The book of Psalm, chapter eight, verse three states, *"When I consider thy heavens, the work of thy fingers, the moon and the stars, which thou hast ordained;"* In two separate passages there are two separate identifiers for heaven. This is further elucidated, *"Behold, the heaven and the heaven of heavens is the Lord's thy God..."*[130] Therefore, there are two heavens: 1) heaven of heavens: the habitation of God and, 2) heaven: the abiding place of the created sun, moon and stars. With this understanding the verse in Nehemiah helps to further expand the understanding of host, *"Thou, even thou, art Lord alone; thou hast made heaven, the heaven of heavens, with all their host..."*[131] There are 'host' related to heaven AND host related to the heaven of heavens. Several biblical passages describe 'host' without necessarily clear understanding of whether the 'host' are attributed to heaven or the heaven of heavens.

Host relevant to the heaven of heavens are created entities shown to be associated with the throne of God. There are two distinct entities of this nature described in scripture; Cheribums and Seraphims.

Seraphim are one of the created entities of the heavenly realm and associated with the heaven of heavens. This entity is described in verses two,six and seven of chapter six in the biblical book of Isaiah, *"Above it stood the seraphims: each one had six wings; with twain he covered his face, and with twain he covered his feet, and with twain he did fly (v. 2)...Then flew one of the seraphims unto me, having a live coal in*

---

129 Genesis 1:8
130 Deuteronomy 10:14
131 Nehemiah 9:6

his hand, which he had taken with the tongs from off the altar: (v. 6) And he laid it upon my mouth, and said, Lo... *(v. 7)"* Although this is a vision Isaiah experienced and was recorded, it shows the physical nature of this entity: the seraphims, 1) had six wings, 2) had feet, 3) had faces, 4) had hands, 5) interacted with man, and 6) communicated with man.

In the biblical book of Ezekiel another distinct type of living creature, Cheribums are described. In chapter one of Ezekiel it is written, *"Also out of the midst there of came the likeness of four living creatures. And this was their appearance; they had the likeness of a man. (v. 5) And every one had four faces, and every one had four wings. (v. 6) And their feet were straight feet; and the sole of their feet was like the sole of a calf's foot: and they sparkled like the colour of burnished brass. (v. 7) And they had the hands of a man under their wings on their four sides; and they four had their faces and their wings. (v. 8) Their wings were joined one to another; they turned not when the went; they went every one straight forward. (v. 9) As for the likeness of their faces, they four had the face of a man, and the face of a lion, on the right side: and they four had the face of an ox on the left side; they four also had the face of an eagle." (v. 10) Thus were their faces: and their wings of every one were joined one to another, and two covered their bodies (v. 11)*[132] In the tenth chapter of the book of Ezekiel, Ezekiel describes what appears to be another living creature but it is in fact a description of the features of the wheel withing a wheel. In verse 14 of chapter ten of the book of Ezekiel it states, *"And every one had four faces: the first face was the face of a cherub, and the second faces the*

---

132 Ezekiel 1:5-11

face of a man, and the third face of a lion, and the fourth face of an eagle." After describing the features of the wheel within a wheel Ezekiel in chapter 10 then refocuses on the cherubims, *"...and I knew that they were the cherubims."*[133] Ezekiel affirms that a cherub is a living creature and that he had encountered it before, *"...This is the living creature that I saw by the river of Chebar."*[134] and this affirmation is reiterated in verse 20 of chapter ten of Ezekiel, *"This is the living creature that I saw under the God of Israel by the river of Chebar."* The living creature seen when Ezekiel was by the river Chebar is identified as a cherub.

Host relevant to the heaven of heavens are created entities shown to be associated with the throne of God. The seraphims described in Isaiah and the cheribums described in Ezekiel are such examples of created entities associated with the throne of God. For example regarding seraphims, the passage in chapter six of Isaiah related above has the lead in verse, verse one, stating, *"...I saw also the Lord sitting upon a a throne..."* Sometimes the likeness of these entities are portrayed on earthly vessels as copies of heavenly realities. One particular host attributed of this nature is the cherub. This entity, the cherub had its likeness fashioned to the mercy seat residing in the tabernacle in the wilderness and later residing the the temple in Jerusalem, *"And make one cherub on the one end, and the other cherub on the other end: even on the mercy seat shall ye make the cherubims on the two ends thereof. (v. 19) And the cherubims shall stretch forth their wings on high, covering the mercy seat with their wings and their faces shall look on to another;*

---

133 Ezekiel 10:20
134 Ezekiel 10:15

toward the mercy seat shall the faces of the cherubims be. (v. 20) "[135]

Other than seraphims and cheribims scripturally the only specific non-generalized heavenly entity is the archangel. This term 'archangel' is theologically the highest ranking angel as etymologically derived. Yet, in scripture there is not definitive evidence as to whether an archangel and angels in general are part of the host of heaven of heavens or the host of the heaven.

Biblically, angles are messengers of God. They are perceived to appear in human likeness when they interact with man on earth, For example, *"And there came two angels to Sodom at even; and Lot sat in the gate of Sodom: and Lot seeing them rose up to meet them;..."* [136]: *"And Jacob went on his way, and the angels of God met him..."*[137]; *"But Mary stood without the sepulchre weeping: and as she wept, she stooped down, and looked into the sepulchre, (v. 11) And seeth two angels in white sitting, the one at the head, and the other at the feet, where the body of Jesus had lain. (v.12) And they say unto her, Woman, why weepest thou?...(v. 13)"*[138] In each of these examples the form of the angel is human-likeness in that there was no fear of their appearance, yet, there was an understanding of their appearance that they were identifiable as angels. However, this perception of angels specifically appearing in human likeness may not be entirely accurate. At this juncture a deviation from strict scriptural referencing is required.

---

135 Exodus 25:19-20
136 Genesis 19:1
137 Genesis 32:1
138 John 20:11-13

Throughout history and various cultures there exist stories of man's interaction with beings of various characteristics. Some human characteristics and some non-human characteristics. Sometimes these beings are viewed as belligerent and even nefarious towards humans. Such examples are demons in christian oriented cultures, and jinn in Islamic and Arabic culture. Many cultures also have traditions and stories of earthly human interactions with what is described as sky people or star people.

Considering the host of heaven as reference to all created beings throughout the universe, excepting humankind provides an understanding to the existence of angels, demons, jinn and other forms of created beings some of which may be currently referenced as space aliens or extraterrestrials. According to current scientific literature, based upon the Hubble Telescope and James Webb Space Telescope there are an estimated 100 billion to 200 billion galaxies in the known universe. When considering the amount of stars per universe; there is an estimated 200 billion trillion stars – or 200 sextrillion stars – or 2×10^{23} stars in the universe.

While scientific inquiry is at odds of life, or more specifically intelligent life or advanced civilizations: in creation this inquiry is moot as the host of heaven is identified as existing. In relation to the scientific inquiry, ufologists have for decades been attempting to verify the existence of some of the host of heaven. Based on purported witnesses and physical encounters there appears to be at least three distinct physical identifiable beings: 1) human-like entities that appear entirely human in appearance, 2) small in stature, large headed, large eyed humanoids, 3) humanoid entities with features of a reptile which are often referred to as reptilian.

The beings identified in 1) above are often referred to as the Nordic as they are described as tall, fair-skinned, blonde hair and fully human in appearance. They are believed to come from the Pleiades star system. The beings identified in 2) above are often referred to as the Grays. These are the beings most often portrayed in science fiction as aliens and are sometimes referred to as the Roswell Grays as they are the life forms reportedly to be retrieved from the 1947 UFO crash in Roswell, New Mexico, United States of America. The Grays are believed to be from the Zeta Reticuli star. The beings identified in 3) above are believed to be nefarious. There are several opinions as to their place of origination. Some believe they come from someplace in the Draco constellation and others opine that they originate from somewhere in the Orion constellation.

While these three species of beings are the most common referenced as reported interactions with humans on earth; a paper in published in *The Astrophysical Journal* estimates there could be as many as 42,777 communicating extraterrestrial intelligent civilizations in our Milky Way galaxy.[139] Thus, within the realm of possibility interaction between humans on earth and beings from the cosmos exists.

In regard to Unidentified Flying Objects (UFO) or

---

139 https://www.forbes.com/sites/jamiecartereurope/2022/05/30/there-are-between-111-and-42777-intelligent-alien-civilizations-in-our-galaxy-say-scientists/

the more recently adopted term Unidentified Anomalous Phenomena (UAP) these apparent physical craft have been observed throughout history. In the Bible in the book of Ezekiel there is a description of an object that is sometimes believed to be a living being but, on close examination of the text it has the characteristics of a vehicle. *"And when I looked, behold the four wheels by the cherubims, one wheel by one cherub, and another wheel by another cherub: and the appearance of the wheels was the colour of a beryl stone. (v. 9) And as for their appearances, they four had one likeness, as if a wheel had been in the midst of a wheel. (v. 10)... As for the wheels, it was cried unto them in my hearing, O wheel. (v. 13) And every one had four faces: the first face was the face of a cherub, and the second face was the face of a man, and the third face of a lion, and the fourth face of an eagle. (v. 14)... And when the cherubims went, the wheels went by them: and when the cherubims lifted up their wings to mount up from the earth, the wheels also turned not from beside them. (v. 16) When they stood, these stood; and when they were lifted up, these lifted up themselves also: for the spirit of the living creatures was in them. (v. 17)"*[140] The cherub and cherubims referenced in the previous passage are living beings. The wheel within the wheel is a physical object controlled by the living beings – 'for the spirit of the living creatures was in them." This tends to indicate an object similar to what we may currently viewed as and described as a drone. While scientific consensus at the present disregards observations of physical flying objects as impossible due to their flight characteristics being incompatible with known physics: this scientific consensus is being challenged because of new

---

140 Ezekiel 10: 9,10,13,14,16,17

observations with corroborating electronic, radar and other advanced telemetry measures and multiplicity of simultaneous sensing arrays.

On November 13, 2024 the United States House Oversight Committee held a hearing regarding Unidentified Anomalous Phenomena. This hearing was a follow-up hearing to one held earlier in the year because UAPs have increasingly been threatening United States airspace and those observing such incursions were being silenced on speaking publicly. While historically such hearings don't usually reveal specifics; this hearing did. Luis Elizondo, former Director of the Pentagon's Advanced Aerospace Threat Identification Program (AATIP) was a sworn witness before the Committee. Upon being questioned by Committee Chairwoman, Anna Paulina Luna about craft being piloted by non-human biologics, Mr. Elizondo responded that they seem to anticipate military maneuvers and he had seen highly sourced emails wherein navy officers referred to their assets as being 'stalked'. Chairwoman Luna recounted that at the previous hearing witness Grusch[141] referred to beings piloting craft as interdimensional beings. Mr. Elizondo deferred classifying beings but responded that based on observable characteristics of instantaneous acceleration the human body can withstand force of about 9 G before negative physiological outcomes (blackout, redout, death), The most capable aircraft the F-16 can withstand 17-18 G before structural failure, vehicles observed during his time at AATIP were

141 David Charles Grusch – former National Reconnaissance Office representative on the Defense Department's Unidentified Aerial Phenomena Task Force whistleblower

performing in excess of 1,000, 2,000, 3,000 G.[142] Based on this reply Chairwoman Luna asked if then were assumed to be "living craft" to which Mr. Elizondo initiated a discourse refusing to define what is living and expresses variances of the identification of life.

There have been reports since the early 1950's of crashed extraterrestrial craft wherein bodies of occupants of the craft have been retrieved. These retrieved bodies were often referred to as EBE – Extraterrestrial Biological Entities. Testimonies before congress in the current time frame about craft and biological retrievals appear to be somewhat consistent with reports from the 1950s.

From ancient times of stories of sky people and angels to present stories of flying craft and extraterrestrial biological entities there is the understanding that – something is out there. There is something out there: in verse one of chapter two of the book of Genesis in the Bible it is stated, *"Thus the heavens and the earth were finished, and all the host of them."* - <u>And all the host of them</u>.

Nikola Tesla a famous twentieth century engineer, physicist and inventor is reported to have said, "If you want to find the secrets of the universe, think in terms of energy, frequency, and vibration." Everything in the universe has energy, frequency and vibration. Energy,

---

142 U.S House of Representatives. Unidentified Anomalous Phenomena: Exposing the Truth – November 13, 2024 – 11:30 am – 2154 Rayburn House Office Building – YouTube video coverage: timestamp [1:41:57 – 1:43:33]. G-force is gravitational force equivalent of a measurement of acceleration.

frequency and vibration are fundamentally interconnected. The amount and type of energy reflects to the vibration and frequency.

Each subatomic particle has a particular frequency; each element has a particular frequency, each molecule has a particular frequency; each planet has a particular frequency; each sun has a particular frequency; each created being – man, beast, host of heaven – has a particular frequency; sub-components of beings, such as individual cells have a particular frequency. For example, the resonate or natural frequency for atomic gold with an isotope of 197 is 8.563 MHz (megahertz) while the resonate or natural frequency for atomic carbon with an isotope of 13 is 125.721 MHz and the resonate or natural frequency for atomic hydrogen with an isotope of 1 is 500 MHz.[143] The resonate or natural frequency for the human body as a whole has not been definitively quantified but ranges between 3 Hz and 10 Hz (hertz). A hertz is a vibration or oscillation of one cycle per second and a megahertz is a vibration or oscillation of one million cycles per second.

There is a particular frequency assumed and attributed to the cosmos as a whole. And there is even assertions that there is a frequency of God.

A thought exercise to contemplate: Imagine you are sitting at the beach watching the waves lap at the beach,

---

143This data obtained from the NMR periodic table: https://imserc.northwestern.edu/guide/eNMR/chem/NMRnuclei.html. The resonate frequency is established within a magnetic field of 11.744T (Tesla – a measurement of magnetic field strength). For reference, the earth has a magnetic field strength of 50 microtesla.

then focusing your attention at the waves when the wind begins to blow creating new waves. The initial wave action had a constant observable pattern, now that pattern has not been changed by the wind but overlaid by a different pattern of waves caused by the wind. Added to this is a passing boat that introduces a new wave pattern. The initial wave pattern did not change, the wind induced wave pattern did not change. Add to this a new wave pattern created by a dolphin swimming nearby. The initial wave pattern did not change, the wind induced wave pattern did not change, the wave pattern caused by the boat did not change. Now consider that a pelican dove into the water for a fish. The initial wave pattern did not change, the wind induced wave pattern did not change, the wave pattern created by the boat did not change, the wave pattern created by the swimming dolphin did not change. You then begin pondering how all of these different wave patterns could co-exist simultaneously.

This thought exercise is but a simplified exercise relative to creation. The complexity, diversity and stability of creation is astounding. The existence of and interactions of quantum fields; the existence and interactions of the quantum particles[144]: up quark, down quark, charm quark, strange quark, top quark, bottom quark, electron lepton, electron neutrino lepton, muon lepton, muon lepton neutrino, tau lepton, tau neutrino lepton, photon boson, w boson, z boson, boson gluons (8), boson graviton, higgs boson; existence and interaction of elements; existence and interactions of atomic isotopes; existence, decay and interactions atomic isotopes; existence and interactions of molecules;

144 Quantum particles theorized under the current Standard Model.

existence of interactions of compounds; existence and interactions of proteins; existence and interaction of DNA; existence and interaction of mitochondrion; existence and interactions of cells; existence and interactions of life forms.

Upon contemplating all of these fundamental physical aspects of creation such contemplation of creation is only partial. Considering the sovereignty of God changes the point of view.

Within the sovereignty of God, the serpent experienced a change to his being. The affect this has on creation needs to be assessed.

Within the sovereignty of God, sin changed initial creation. The affect sin has on creation needs to be assessed.

Within the sovereignty of God, the soul affects creation. This needs to be assessed.

Within the sovereignty of God, spirits affect creation. This needs to be assessed.

Within the sovereignty of God, salvation affects creation. This needs to be assessed.

Within the sovereignty of God, seraphim & such affects creation. This needs to be assessed.

Creation is so much more than the physical aspects that science seeks to identify and understand. The 'butterfly effect' is a way of describing that everything in the universe is interconnected. Derived from chaos theory, the butterfly theory expounds that hypothetically a flap of a butterfly's wing could cause a hurricane. Yet, the interconnectedness of creation is so much more complex and extends from the beginning of creation to eternity. For example, a man's (human being) thoughts are scientifically in the physical realm ascertained to be an energy induced wave. How though does this wave affect it surroundings and ultimately affect creation as a whole?

The aspects of creation: sovereignty, serpent, sin, soul, spirit, salvation, seraphim & such (and so many other things) all meld with the scientific observable and theoretical aspects to truly signify the sublime of creation. Sublime in definition as: lofty, grand, or exalted in thought, expression, or manner; of outstanding spiritual, intellectual, or moral worth; tending to inspire awe usually because of elevated quality (as of beauty, nobility, or grandeur) or transcendent excellence.[145]

---

145 Sublime definition: https://www.merriam-webster.com/dictionary/sublime

Thomas W. Guthrie

CHAPTER 11
CONTROLLING TIMELINE

Scripture is very explicit in the timeline of creation...on day 'X' this happened...on day 'Y' this happened. The breadth of creation is detailed in chapter 1 of Genesis of the Old Testament. Certain details regarding the creation of Adam are then further clarified in chapter 2 of Genesis. Starting in verse 4 of chapter 2, of Genesis is in context a parenthetical addendum within chapter 1.

The major components of creation and in essence, the universe: time, heaven, earth, sun, moon, stars, light, darkness, day, night and man are explained in chapter 1 of Genesis. This creation is enveloped in seven literal 24 hour days. The text and context of ordinal nature – first day, second day, third day, fourth day, fifth day, sixth day and seventh day are foundational for such chronology. Ordinal numbers indicate rank and order, and not quantity. Such ordinal emphasis within creation is compounded when considering Adam.

The order of creation in Genesis, chapter 1:

"In the beginning God created the heaven and earth. (v. 1) And the earth was without form, and void; and darkness was upon the face of the deep. And the Spirit of God moved upon the face of the waters. (v. 2) And God said, Let there be light: and there was light. (v. 3) And God saw the light, that it was good: and God divided the light from the darkness. (v. 4) And God called the light Day, and the darkness he called Night. And the evening

and the morning were the first day. (v. 5)

And God said, Let there be a firmament in the midst of the waters, and let it divide the waters from the waters. (v. 6) And God made the firmament, and divided the waters which were under the firmament from the waters which were above the firmament: and it was so. (v. 7) And God called the firmament Heaven. And the evening and the morning were the second day. (v.8)

And God said, Let the waters under the heaven be gathered together unto one place, and let the dry land appear: and it was so. (v. 9) And God called the dry land Earth; and the gathering together of the waters called the Seas: and God saw that it was good. (v. 10) And God said, Let the earth bring for grass, the herb yielding seed, and the fruit tree yielding fruit after his kind, whose seed is in itself, upon the earth: and it was so. (v. 11) And the earth brought fort grass, and herb yielding seed after his kind, and the tree yielding fruit, whose seed was int itself, after his kind: and God saw that it was good. (v. 12) And the evening and the morning were the third day. (v. 13)

And God said, Let there be lights in the firmament of the heaven to divide the day from the night; and let them be for signs, and for seasons, and for day, and years: (v. 14) And let them be for lights in the firmament of the heaven to give light upon the earth: and it was so. (v. 15) And God made two great lights; the greater light to rule the day, and the lesser light to rule the night: he made the stars also. (v.16) And God set them in the firmament of the heaven to give light upon the earth, (v. 17) And to rule over the day and over the night, and to divide the light from the darkness: and God saw that it was good. (v. 18)

And the evening and the morning were the fourth day. (v. 19)

And God said, Let the waters bring forth abundantly the moving creature that hath life, and fowl that may fly above the earth in the open firmament of heaven. (v. 20) And God created great whales, and every living creature that moveth, which the waters brought forth abundantly, after their kind, and every winged fowl after his kind: and God saw that it was good. (v. 21) And God blessed them, saying, Be fruitful, and multiply, and fill the waters in the seas, and let fowl multiply in the earth. (v. 22) And the evening and the morning were the fifth day. (v. 23)

And God said, Let the earth bring forth the living creature after his kind, cattle, and creeping thing, and beast of the earth after his kind: and it was so. (v. 24) And God made the beast of the earth after his kind, and cattle after their kind, and every thing that creepeth upon the earth after his kind: and God saw that it was good. (v. 25) And God said, Let us make man in our image, after our likeness: and let them have dominion over the fish of the sea, and over all the earth, and over every creeping thing that creepeth upon the earth. (v. 26) So God created man in his own image, in the image of God created he him; male and female created he them. (v. 27) And God blessed them, and God said unto them, Be fruitful, and multiply, and replenish the earth, and subdue it: and have dominion over the fish of the sea, and over the fowl of the air, and over very living thing that moveth upon the earth. (v. 28) And God said, Behold I have given you every herb bearing seed, which is upon the face of all the earth, and every tree, in the which is the fruit of a tree yielding seed; to you it shall be for meat. (v. 29) And to every beast of

127

> *the earth, and to every fowl of the air, and to every thing that creepeth upon the earth, wherein there is life, I have given every green herb for meat: and it was so. (v. 30) And God saw every thing that he had made, and, behold, it was very good. And the evening and the morning were the sixth day. (v. 31)*

Creation is culminated in verses one through three of chapter 2 of Genesis:

> *Thus the heavens and the earth were finished, and all the host of them. (v. 1) And on the seventh day God ended his work which he had made; and he rested on the seventh day from all his work which he had made. (v. 2) And God blessed the seventh day, and sanctified it: because that in it he had rested from all his work which God created and made. (v. 3)*

Thus, evening and morning for six days – 24 hour days - God created, and on the seventh – 24 hour day – he rested. This is exemplified by Adam. Scripture shows when during the six days of creation that Adam was created and details his life span and linage. Within the account of creation Adam is identified as a specific creation. And within history, Jesus of Nazareth, the Messiah identifies Adam as a historical figure. It is via Adam that the specificity of the timeline of creation can be ascertained.

Details of the creation of Adam are found in verses four through seven of chapter 2 of Genesis:

> *These are the generations of the heavens and of the earth when they were created, in the day that the LORD God made the earth and the heavens, (v. 4) And every plant of*

the field before it was in the earth, and every herb of the field before it grew: for the LORD God had not caused it to rain upon the earth, and there was not a man to till the ground. (v. 5) But there went up a mist from the earth, and watered the whole face of the ground. (v. 6) And the LORD God formed man of the dust of the ground, and breathed into his nostrils the breath of life; and man became a living soul. (v. 7)

For clarity regarding this lineage as is defined in scripture later in chapter 2 of Genesis, the Hebrew transliteration of 'Adam' will be used instead of 'man'. Following the text and context of these passages in chapter 2 of Genesis, God created Adam "*...in the day that the LORD God made the earth and the heavens...*" Referring back to details of creation from chapter 1 of Genesis, the day that God made the earth and heavens was on the third day (Chapter 1, verses 9 through 13). According to chapter 1 of Genesis, it was also the third day when the grass, herbs and trees were created. Yet, Chapter 2 of Genesis states that Adam was created prior to creation of the grass, herbs and trees but still within the same day.

So Adam has a definitive beginning on the third day of creation.[146] It is from the lineage of Adam that the peoples of earth are reckoned and the continuing timeline of creation is reckoned.

The lineage of Adam begins with the birth of his first sons Cain and Able in Genesis, chapter 4. However, with the murder of Abel by Cain[147] the limited lineage of Adam through his son, Cain is detailed in verses 17 through 26 of chapter 4 of Genesis.

146This seeming conflict with the creation of man on the sixth day in verse 27 of chapter 1 of Genesis will be vetted in a subsequent chapter.

147Genesis, chapter 4, verse 8

Within this limited lineage no timeline or number of years is ascribed to this narrative. But, it is imperative to remember that on the fourth day of creation; time was verified in scripture, *"And God said, Let there be lights in the firmament of the heaven to divide the day from the night; and let them be for signs, and for seasons, and for days and years:"*[148]. Therefore, there is no legitimate opportunity to interpret or postulate a broad or open-ended period of an 'age' outside of seasons, days and years. It is established in creation that the one function of celestial bodies is for the reckoning time.

Starting in chapter 5 of Genesis, with the birth of Seth and lineage through Seth that a timeline via the numbering of years of births and deaths is set forth. In verse three of Genesis, chapter 5 it is stated, *"And Adam lived an hundred and thirty years, and begat a son in his own likeness, after his image; and called his name Seth:"* Verses four and five of chapter 5 of Genesis go on to completed the numbering of years for Adam, *"And the days of Adam after he had begotten Seth were eight hundred years; and he begat sons and daughters: And all the days that Adam lived were nine hundred and thirty years: and he died."* This narrative shows that the numbering of years is consistent in a life of 930 years that was composed of 130 years from Adam's creation to the birth of Seth and 800 years after Seth's birth until Adam died (130 + 800 = 930). The narrative also confirms that Adam experienced 'the way of man' – death. While it is difficult to fathom a life span of 930 years from a current perspective from the year reckoned as 2021 AD (CE): such lack of the ability to fathom the lengthy life span does not negate the narrative.

The time line of creation continues through the lineage of Seth. This entirety of the time line can effectively be segmented

---

148Genesis 1:14

into three periods: 1) a period from the lineage of Adam through the patriarch Abraham[149], 2) a period from the patriarch Abraham to the advent of the Messiah (Christ), 3) a period from the advent of the Messiah to the present – currently reckoned as 2021 AD (CE). The first and third periods are relatively straight forward. And, while the second period has some complexity; it is determinable. It is imperative to remember that the time line of creation is detailed within the Hebrew perspective and overall relative to God's interactions with man.

The first of the three periods is from the lineage of Adam through Seth and continues through to the patriarch Abraham. When viewing the life spans of the patriarchs it is common to focus on the life span of each individual. Although this is common, it is errant to use the overall life span of a individual to determine the time line of creation. The proper focus is on the age of the individual when the referenced son is begotten. Otherwise, there is an overlap of timing that skews the final numbering of years. Beginning with Adam: a) Adam was 130 years old when he begat Seth [Genesis 5:3], b) Seth was 105 years old when he begat Enos [Genesis 5:6], c) Enos was 90 years old when he begat Cainan [Genesis 5:9], d) Cainan was 70 years old when he begat Mahalaleel [Genesis 5:12], e) Mahalaleel was 65 years old when he begat Jared [Genesis 5:15], f) Jared was 162 years old when he begat Enoch [Genesis 5:18], g) Enoch was 65 years old when he begat Methuselah [Genesis 5:21], h) Methuselah was 187 years old when he begat Lamech [Genesis 5:25], i) Lamech was 182 years old when he begat Noah [Genesis 5:28,29], j) Noah was 500 years old and begat Shem, Ham and Japeth [Genesis 5:32], k) Shem was 100 years old when he begat Arphaxad [Genesis 11:10], l) Arphaxad

---

149Although the scripture lineage delineates "Abram", the changed name of "Abraham" (changed in scripture from Abram to Abraham) is used to be readily recognizable.

was 53 years old when he begat Salah [Genesis 11:12], m) Salah was 30 years old when he begat Eber [Genesis 11:14], n) Eber was 34 years old when he begat Peleg [Genesis 11:16], o) Peleg was 30 years old when he begat Reu [Genesis 11:18], p) Reu was 32 years old when he begat Serug [Genesis 11:20], q) Serug was 30 years old when he begat Nahor [Genesis 11:22], r) Nahor was 29 years old when he begat Terah [Genesis 11:24], s) Terah was 70 years old when he begat Abram.

The addition of this time spans – 'a' through 's' equals the time span related to the first period of the time line of creation:

[a) 130 + b) 105 + c) 90 + d) 70 + e) 65 + f) 162 + g) 65 + h) 187 + i) 182 + j) 500 + k) 100 + l) 53 + m) 30 + n) 34 + o) 30 + p) 32 + q) 30 + r) 29 + s) 70 = 1,964]

The first period of the time line of creation is easily tabulated. According to scripture the time from the creation of Adam to the birth of Abraham was 1,964 years.

The complexity of arriving at tabulated years for the second period still relies on the perspective of events within Hebrew history. With all of the necessary elements being present in scripture. The first two points are, as earlier tabulation, straight forward in the birth of Abraham's son, Isaac and the birth of Issac's son, Jacob. Abraham had his son Issac when Abraham was 100 years old.[150] Isaac was 60 years old when his son Jacob was born.[151]

Jacob himself had 12 sons. Jacob's son, Joseph is the last individual from which tabulating a timeline is applicable. Although not directly from the birth of a specific son, 'X' begat

150Genesis 17:17, 21
151Genesis 25:26

'Y' at 'Z' years old, but, from events associated in the life of Joseph which have time referencing.

The narrative regarding Joseph begins in chapter 37 of Genesis. Joseph, a son of Jacob residing in Canaan was sold into slavery by his brothers at the age of 17.[152] Although it is not directly stated how old Jacob was when Joseph was born, scripture records that Joseph was born in Jacob's old age.[153] In Genesis, chapter 37, verse 28 it is recorded that Midianite merchantmen retrieved Joseph from a pit in which his brothers had placed him and then sold Joseph to Ishmeelites who took Joseph to Egypt.

At the age of 30, Joseph through a series of events finds his way standing before Pharaoh king of Egypt.[154] Joseph at this time within the court of Pharaoh was put in charge of gathering food for storage for a period of seven years in order to prepare for drought. A drought and famine had been prophesied for a period of seven years after a period of seven years of fruitful times. It is recorded in Genesis, chapter 41, verse 54 that the drought and famine did manifest. The drought and famine was not in Egypt only but affected Joseph's father's family in Canaan. Ultimately, the sons of Jacob came to Egypt seeking provisions to survive. Joseph's brothers did not initially recognize him. But, Joseph revealed himself to his brothers with five years remaining in the drought and famine.[155] According to Genesis, chapter 45, verses 17 and 18 it was at this time that Pharaoh invited Joseph's father and brothers along with their families to reside in Egypt. When Jacob came to Egypt, Pharaoh inquired Jacob of his age and in verse 9 of chapter 47 of Genesis Jacob

---

152Genesis 37:1,2
153Genesis 37:3
154Genesis 41:46
155Genesis 45:11

replies, *"The days of the years of my pilgrimage are an hundred and thirty years..."*

In order to calculate the age of Jacob at the time of Joseph's birth for the purpose of using that age to add to the tabulation of the timeline of creation; the times associated with the story of Joseph are required to be used. Joseph was sold at the age of 17 and stood before Pharaoh at the age of 30.Therefore, Joseph was in Egypt 13 years by the time the seven years of fruitfulness began. There were 7 years of fruitfulness. Joseph revealed himself to his brothers in the 2nd year of drought and famine and at which time Jacob and his sons and their families came into Egypt. The total amount of time associated with Joseph in Egypt before Jacob arrived is 22 years (13 + 7 + 2 = 22). Now take into account that Joseph was 17 years old when he entered into Egypt through slavery. Accordingly, Joseph was 39 years (17 + 22 = 39) old when Jacob rejoined Joseph. And, Jacob declared to Pharaoh at that time that he was 130 years old. When subtracting Joseph's age from Jacob's age we find that Jacob was 91 years old (130 − 39 = 91) when Jacob begat Joseph thus confirming that Jacob was of an old age when he begat Joseph.

However, the relevance of Jacob's age at the birth of Joseph is only a partial component of tabulation of the timeline of creation. Since the next event of import is when the Hebrew people began their residency in Egypt under Pharaoh; Jacob's age at the birth of Joseph is added to the time Jacob was away from Joseph during the time Joseph was in Egypt. This time equals the amount of years Jacob told Pharaoh he had been alive at the time of his arrival in Egypt. Thus, 130 years is added to the tabulation of the timeline of creation.

The next way point in tabulating the timeline of creation is the time the Hebrews spent in the land of Goshen in Egypt prior to the exodus. This period is associated with the figure Moses and the plagues wrought upon Egypt and Pharaoh in order to secure the release of the Hebrews from Egyptian bondage. Tabulating this way point is much simpler than the story of Joseph. The book of Exodus at verse 41 of chapter 12 records, *"And it came to pass at the end of the four hundred and thirty years, even the selfsame day it came to pass, that all the hosts of the Lord went out from the land of Egypt."* 430 years is our next figure for tabulation.

Upon exiting from Egypt the Hebrews began a journey back to the land of Canaan. This journey was not as simple as the arrival into Egypt by Jacob, his sons and their families 430 years earlier. Scripture reports that there was contention by the people against Moses and his leadership. Yet it was during this period that the ritual worship elements were established that are generally referred to as the Mosaic Law. Upon reaching Canaan after this relatively short journey; rebellion rose within the people due to negative reports about the land of Canaan which the Hebrew people were about to enter. This rebellion caused a judgment against them that caused them to not enter into Canaan and wander in the wilderness for 40 years.[156] After this infamous wandering in the wilderness for 40 years, the Hebrew people did once again inter into the land of Canaan under the leadership of a man named Joshua. Although the Hebrew people had always had historical interaction with other people groups; upon entering Canaan such interactions became more documented. The Hebrew people entering into Canaan resulted in multiple battles and wars with various groups such as the Canaanites, Hittites, Hivites, Perizzites, Girgashites, Amorites and the Jebusites.[157] This interaction took place over a long period of time. After the passing of Joshua, leadership among the Hebrews

---

156Numbers 14: 28-33
157Joshua 3:10

was nebulous as a whole people with the function of leadership residing with in each 'tribe' or group of people based on the lineage of the sons of Jacob. Initially, any leadership of the people as a whole was undertaken by 'judges'. However, in a desire to be like the surrounding people groups the Hebrew people pursued anointing kings to rule over them. The first was king Saul, the second was king David and the third was king Solomon. Although Saul was the first king of the Hebrew people, the lineage through David is the route for continuing to follow the timeline of creation. King David's son, Solomon being the first heir to the throne through the lineage of David. For broad reference, king Solomon is the man reputed for his wisdom and wealth throughout the whole earth.

The next way point for tabulating the time line of creation resides within the kings. There is a definitive time frame set within the reign of Solomon is the book of 1 Kings, chapter 6, verse 1. It states, *"And it came to pass in the four hundred and eightieth year after the children of Israel were come out of the land of Egypt, in the fourth year of Solomon's reign over Israel..."*[158] But, in order to finish the time line of creation over the course of the time of the patriarchs; reversion to the begetting of king David is necessary. In verse 4 of chapter 5 of the book of 2 Samuel is is recorded, *"David was thirty years old when he began to reign, and he reigned for forty years."* Thus by subtracting the age when king David began to reign and subtracting the number of years he reigned and subtracting the four years of reign of king Solomon from the 480 years stated in the book of 1 Kings, chapter 6, verse 1; we find that from the exodus from Egypt by the Hebrew people to the begetting of king David was 406 years $(480 - 40 - 30 - 4 = 406)$.

---

[158]The nation or country of Israel is a national reference to the Hebrew people based on the change of name of Jacob to Israel in Genesis 32:28.

From this point, from king David through the period of the kings the number of years each king reigns will be the method of tabulating the timeline of creation. Of course, the 30 years of king David's life prior to the beginning of his reign is required to eventually be added back to the tabulation.

The period of the kings of the Hebrew people was like most kingdoms throughout history filled with wars, family squabbles, palace intrigue and assassinations. In fact, after the reign of Solomon, king David's son the kingdom of Israel was divided into two distinct kingdom's – Israel and Judah. Israel was identified as the "Northern Kingdom" consisting of ten tribes reckoned according to the sons of Jacob (Israel). Judah was identified as the "Southern Kingdom" consisting of two tribes reckoned according to the sons of Jacob (Israel). These two tribes being Judah and Benjamin.

It is the Southern Kingdom of Judah which will be followed. As many of the kings among the two kingdoms overlap, following both is confusing and unnecessary. There are many interesting kings, wars and events associated with each king of Judah however, refocusing to the elemental time frame for each reign is prudent. As with the 'begetting' of the patriarchs; such listings might not prove to be exciting, but: a') David reigned 40 years [2 Samuel 5:4]; b') Solomon reigned 40 years [1 Kings 11:42 & 2 Chronicles 9:40]; c') Rehoboam reigned 17 years [1 Kings 11:43 & 14:31]; d') Abijah reigned 3 years [1 Kings 14:31 & 15:8]; e') Asa reigned 41 years [1 Kings 15:8-24]; f') Jehoshaphat reigned 25 years [1 Kings 22:41-50]; g') Jehoram (Joram) reigned 8 years [2 king 8:16-24]; h') Ahaziah reigned 1 year [2 Kings 8:24 & 9:29]; i') Athaliah (Queen) reigned 6 years [2 Kings 11:1:20]; j') Joash (Jehoash) reigned 40 years [2 Kings 11:1 & 12:21]; k') Amaziah reigned 29 years [2 Kings 14:1-20]; l') Uzziah (Azariah) reigned 52 years [2 kings 15:1-7]; m') Jotham reigned 18 years [2 Kings 15:32-38]; n') reigned 19 years [2 Kings 16:1-20]; o') Hezekiah reigned 19 years [2 Kings 18: & 20:21]; p') Manasseh reigned 55 years [2 Kings 21:1-18]; q') Amon reigned 2 years [2 Kings 21:19-26]; r') Josiah (Josias) reigned 31 years [2 Kings 22:1 & 23:30]; s') Jehoahaz (Joahaz) reigned 3 months [2 Kings 23:31-33]; t') Jehoiakim reigned 11 years [2 Kings 23:34 & 24:5]; u') Jehoiachin reigned 3 months [2 Kings 24:6-16] and v') Zedekiah reigned 11 years [2 Kings 24:17 & 25:30].[159]

The addition of this time spans – 'a" through 's" equals the time span related to the time of the Hebrew kings within the time line of creation:

[a') 40 + b') 40 + c') 17 + d') 3 + e') 41 + f') 25 + g') 8 + h') 1 + i') 6 + j') 40 + k') 29 + l') 52 + m') 18 + n') 19 + o') 19 + p') 55 + q') 2 + r') 31 + s') $1/4^{th}$ + 't) 11 + u') $1/4^{th}$ + v') 11 = 437.50]

The times of the kings are sequentially one after the other without any intermittent pauses or gaps. Therefore, the time from the start of the reign of king David through end of the reign of king Zedekiah is 437.50 years.

So, the timeline of creation from Adam through king Zedekiah is 3,558.50 years: Adam to Abram (Abraham) – 1,964 years; Abraham to the Hebrew people's residency in the land of Egypt – 291 years; The years of the Hebrew people's residency in the land of Egypt – 430 years; Exodus of the Hebrew people from the land of Egypt to the reign of king David – 436 years; Reign of king David through the reign of king Zedekiah – 437.50 years.

159Listing and referencing for Rehoboam through Zedekiah obtained from chart of the kings located at https://www.conformingtojesus.com

Scripure at the point of the end of the reign of king Zedekiah ceases to provide direct year by year referencing for tabulating the timeline of creation. During the reign of Jehoiakim, king Nebuchadnezzar of Babylon beseiged Jerusalem and took many of its inhabitants captive into Babylon.[160] Additional attacks by Nebuchadnezzar and subsequent periods of captivity of the inhabitants of Jerusalem and Judah into Babylon also took place under the reigns of Jehoiachin and Zedekiah.[161] These events interrupted the continuity of the narrative of the Hebrew people. Although scripture still deals with the the Hebrew people; the focus changes from their journey from the beginning of time to their sufferings in history. None the less, the timeline of creation is still rooted in and progresses through them.

While year to year tablution of the timeline of creation within scripture ceases, scripture provides one additional 'marker' for continued tabulation. In the book of Jeremiah, chapter 25, verse 11 it was prophsied, *"And this whole land shall be a desolation, and an astonishment; and these nations shall serve the king of Babylon seventy years."* In context, this is a prophecy regarding the seige of Jerusalem under king Nebuchadnezzar and the captivity of the Hebrew people. The fulfillment of this prophecy is referenced both in the book of 2 Chronicles, chapter 36, verses 21 and 22 (*"And them that had escaped from the sword carried he away to Babylon; where they were servants to him and his sons until the reign of the kingdom of Persia: (v. 20) To fulfil the word of the Lord by the mouth of Jeremiah, until the land had enjoyed her sabbaths: for as long as she lay desolate she kept sabbath, to fulfil threescore and ten years. (v. 21)"*) and the book of Daniel, chapter 9, verse 2 (*"In the first year of his reign I Daniel understood by books the*

160Daniel 1:1
1612 Chronicles 36:9-20

number of the years, whereof the word of the Lord came to Jeremiah the prophet, that he would accomplish seventy years in the desolation of Jerusalem."). Accordingly, another 70 years could be added to the timeline of creation. Yet after this 70 years, the direct tabulation of the timeline of creation from scripture ceases.

All is not lost though in following the timeline of creation. At this point in history the interaction of the Hebrew people with recogized world powers began to be well documented. Records of the Babylonians, the Persian, the Greeks and the Romans all document their interactions with the Hebrew people. And, the documentation of the Hebrew people with these world powers allow the timeline of creation to be carried forth in history. Every culture and civilization has accorded to their unique setting and circumstance a method for keeping years. Effectively, time keeping within a civilization resides within tracking kings, their length of reign and their successor. This methology as such has been seen within the Hebrew people with the reign of kings from David through Zedekiah. How are years asscertained when organized leadership of a civilization ceases? Fortunately, there is a conquering civilization or a defining event that allows a continuation of tracking years. Such is the case with the Hebrew people.

With the Babylonian kingdom decimating and enslaving the kingdom of Judah; the timeline of creation is tracked through the kings of Bablyon and then through successive conquerors – the Persians, the Greeks, the Romans, etc. In the second century of the Christian era (AD (CE)), Claudius Ptolemy an astronomer and geographer using documents within the library of Alexandria, Egypt compiled and cross-referenced a list of kings based on the kings of Egypt. The document he created in known as Ptolemy's Canon of Kings. And, with Claudius Ptolemy being an astronomer, the list of kings is cross-referenced with documented astronomical events such as eclipses.

Ptolemy's Canon of Kings has been used as the historical standard for understanding the succession on kings through the various civilization of the Middle East until the time of the advent of Jesus of Nazareth, commonly referred to as Jesus Christ, or Christ. According to Livis.org, "This brief document, which is based on astronomical information from ancient Babylon, is still the backbone of the chronology of the ancient Near East. Its essential correctness has been corroborated by the Uruk King List, the *Astronomical Diaries*, and Egyptian data (like date papyri)."[62] Based on Ptolemy's Canon of Kings the time period from king Jehoiachin's captivity by Nebuchadnezzar until the time of Christ is 587 years. Therefore, 587 years becomes the next point of tabulation in the timeline of creation.

The second period of the timeline of creation, a period from the patriarch Abraham to the advent of the Messiah (Christ):

Abraham to the Hebrew people's residing in Egypt – 291 years
Hebrew people in Egypt to their exodus – 430 years
Period from the exodus to the reign of king David – 436 years
Reign from king David through king Zedekiah – 437.50 years
Period without kings until the Messiah (Christ) – 587 years

Thus, the second period in the timeline of creation is composed of 2,181.50 years.

The third and final period for ascertaing the of the timeline of creation is the period from the time of Christ to the present. As with the time of the Hebrew people without a king, using the

---

162https://www.livius.org/articles/concept/ptolemys-canon/

advent of Christ as a focal point requires tracking reigns of kings and other signifigent events. Over the course of the time from the advent of Christ there have been several calendar changes and modifications. For example, in 1582 AD (CE) the Julian Calendar which was authorized by Julius Caesar in 45 BC (BCE) was replaced in Europe by the Gregorian calendar in 1582 AD (CE). Thus, in terms of present day accounting the current year is 2024 AD (CE). The referencing of "AD" is abbreviation for "Anno Domini" a Latin term for "in the year of our Lord" – thus referencing Christ. The abbreviation "CE" is reference to the year from day one on the Gregorian calendar but, retaining a neutral – non-Christ or non-Christian reference. However, since the Gregorian calendar is based on the advent of Christ, AD (CE) or BC [Before Christ] (BCE) [Before Commone Era] are still Christ associated.

With the timeline of creation arriving at the present, 2024 becomes the next point in tabulation and allows the culumination of the tabluation. The three periods as earlier referenced:

1) a period from the lineage of Adam through the patriarch Abraham - 1,964 years
2) a period from the patriarch Abraham to the advent of the Messiah - 2,181.50 years
3) a period from the advent of the Messiah to the present – 2024 years

Therefore, the tabulated timeline of creation since creation is 6,169.50 years; or rounded to 6,170 years. However, this tabulated timeline is required to be reduced by four years since the monk Dionysius Exigus who devised the time system since the advent of the Messiah, Jesus of Nazareth erred in the calculation of the death of Herod the Great. Accordingly, the period from the advent of the Messiah to the present is 2020 years and the time from creation to the present is 6,166 years. As a point of reference, the Jewish year for the Gregorian year of 2021 is 5784-5785 [163]. The difference of 381 years between Jewish timekeeping and the tabulation contained herein is inconsequential when comparing it against the secular perceived time since creation of 12-14 billion years.

Yet, when one considers the text of the Bible; re-evaluation of time is substantiated. God gives markers to time. For example, if one contemplates earth geological processes of both evolutionary (slow processes) and cataclysmic (fast processes) billions of years of change would be highly unlikely to retain physical geological landmarks. However, in Genesis, chapter 2 the garden planted by God called Eden details its physical location on earth. Beginning at verse 10 of Chapter 2:

"And a river went out of Eden to water the garden; and from thence it was parted, and became into four heads. (v.10) The name of the first is Pison: that is which compasseth the whole land of Havilah, where there is gold; (v. 11) And the gold of that land is good: there is bdellium and the onyx stone. (v. 12) And the name of the second river is Gihon: the same is it that compasseth the whole land of Ethiopia. (v. 13) And the name of the third river is Hiddekel: that is it which goeth toward the east of Assyria. And the fourth river is Euphrates. (v. 14)"

For anyone with a modicum of geographically understanding, Ethiopia and the Euphrates River are still recognizable as valid locations on the earth. Yet, for the purpose of clarity the Euphrates is the river having its source in the current geopolitical country of Turkey and flowing through the current geopolitical countries of Syria and Iraq emptying into the

---

163 https://www.chabad.org/calendar/view/year.htm

Persian Gulf. Ethiopia, although currently a geopolitical country is an area in northeast Africa whose location has been consistent and acknowledged throughout antiquity. Thus, these two remain observable markers that have not been obliterated by the geological processes of an extended period of time – such as multiple million or billion years. So, within the biblical text there is both a strict discernible timeline from creation and observable markers.

CHAPTER 12
RECONCILIATION

Science makes a determination of the timeline of creation based on multiple theories. Although theories are a basis and necessity for the scientific process, according to the scientific process theories wherein experimental or observable data opposes a theory; the theory is required to be discarded. According to the scientific method there is no "settled science". As advancements are made in experimental methods to verify or disprove theories new theories may be necessary.

An example of this is the evolution of scientific understanding of the atom. Originally, the atom was believed to be a single constituent stand-alone indivisible unit. Later, through additional experimentation it was theorized the the atom was made up of three indivisible units: the nucleus composed of a proton and a neutron and orbiting around the nucleus, the electron. Under current theory and experimentation the atom is composed of multiple quantum particles. And, the electron(s) are not single particle(s) orbiting around a nucleus but a cloud of energy vibrating around the vibrating nucleus.

Cosmology and geology have had and continue to have evolution of theories. But, when it comes to time, age and length of existence since creation there has been very little evolution of theories in the last 100 years. Current scientific consensus believes the age of the universe is about 13.8 billion years old. Current scientific consensus believes the age of the earth is about 4.5 billion years old. But, how are these ages determined?

Regarding the universe, there are two main theories from which this age is derived: the theory of the Big Bang, and the theory of an expanding universe. In simple terms, the Big Bang makes the assumption that all things in the universe were at one point in time existing at a single point (singularity) and some event caused all things to be dispelled instantaneously from the point of singularity. The theory of the expanding universe stipulates that since there was a Big Bang the universe is expanding away from the point of singularity. The age of the universe in then determined extrapolating the rate of expansion backward in time. One experimental method of attempting to validate the theory of an expanding universe is studying redshift. Redshift is an observation that at lengthy distances elements absorption of light waves spectrally shift toward the infared of the light spectrum – hence – redshift. For example, when hydrogen is viewed on the earth in spectrum analysis it has absorption points in the light spectrum at 410 nanometers, 434 nanometers, 486 nanometers and 656 nanometers. When hydrogen is viewed in deep space using spectrum analysis these points of absorption all move toward the infared of the light spectrum. However, in astrophysics there are three recognized types of redshift: doppler redshift caused by movement away from the observer, gravitational redshift caused by light moving from a strong gravitational field, and cosmological redshift caused by the expansion of space. Each of these redshift types subsequently rely on additional theories to differentiate between them.

Although there is some empirical data to support an expanding universe, there is also data that would discount the theory. A study of redshift in nearby stars indicates the invalidates the expanding universe theory.[164] So, although the

164 Kalhor, Bahram & Mehrparvar, Farzaneh & Kalhor, Behnam. (2021).

theory of an expanding universe may be a consensus theory among cosmologists and astrophysicists; there are alternate theories being explored. If through future scientific inquiry the Big Bang theory and the expanding universe theory prove to be untrue, how does such alter perception of understanding of the age of the universe? Is the universe really 13.8 billion years old?

As to determining the age of the earth scientists also rely on theories as well. A key consensus theory related to this is the geological theory of uniformity. This theory states the geological processes on the earth today are the same processes taking place throughout the history of the earth and these processes are gradual and cyclical. The theory recognizes that although cataclysmic events do occur they are within the confines of gradualism. Gradual build up of rock and soil strata, and gradual erosion of strata. Within this theory the age of the earth is assumed to be about 4.5 billion years old. Empirical data for this theory has been established by theories using radiometric dating.

Radiometric dating assumes that certain isotopic elements such as carbon-14, potassium-40, rubidium-87 and uranium-238 are unstable isotopes that undergo radioactive decay while stable isotopes do not undergo radioactive decay. Therefore, a specimen is analyzed for the amount of decayed isotope remaining in proportion to the amount of the stable isotope. For example, carbon-12 is a stable, naturally occurring and predominate isotope of carbon that does not undergo radioactive decay. By measuring the amount of carbon-14 in a specimen and determining its ratio to carbon-12 in the specimen, the age of a specimen can be determined. As each isotope (carbon-14, potassium-40, rubidium-87 and uranium-238) has a different radioactive half-life[165] each is used for different lengths of time.

Unexpected Redshift cf nearby stars. 10.21203/rs.3.rs-234589/v1.
165 Half-life is the time it takes for a quantity of radioactive isotope to fall to

For example, the half-life of carbon-14 is 5,730 years, the half-life of potassium-40 is 1.25 billion years, the half-life of rubidium-87 is 48.8 billion years and the half-life of uranium-238 is 4.468 billion years. Accordingly, each isotope is used for determining age for different periods of time. Carbon-14 is used for determining ages of relatively young items up to an age of 50,000 years old, and usually associated with living organisms. Potassium-40, rubidium-87 and uranium-238 are used for determining the age of rocks.

While this methodology of dating has been standardized and generally accepted in the field of geology, there is empirical data showing the the methodology to be errant, to some degree. In a published paper available on the University of North Carolina website[166] More Bad News for Radiometric Dating wherein the author discusses several studies that analyze the processes associated with magma separation, cooling and crystallization and how different circumstances affect determining its age. One salient quote from the paper provides this insight, "Some information from the book Uranium Geochemistry Mineralogy, Geology provided by Jon Covey gives us evidence that fractionation processes are making radiometric dates much, much to old." Another study[167] regarding the presence of carbon-14 in natural diamonds also suggests errancy in the veracity of radiometric dating. Diamonds are believed to be between one and two billion years old yet carbon-14 which has a half-life of 5,730 years still exists in diamonds. Accordingly, if their age is one to two billion years;

one-half of its initial value.

166 https://www.cs.unc.edu/~plaisted/ce/dating2.html

167 https://answersingenesis.org/geology/carbon-14/radiocarbon-in-diamonds-confirmed/?srsltid=AfmBOoozxiuiNrjTAEUmkzhUF-KbkAm1BG_1PPIAc2NJUWEnEzZCGqvGl. This study also footnotes that the presence of carbon-14 in now being reported in conventional literature

there should be no carbon-14 present in diamonds. Other factors that may affect radiometric dating is the presumption that there is and always has been a constant proportional ratio of stable isotopes to radioactive isotopes, and that radioactive decay rates are not influenced by unusual external events. If either or both of these factors are proven to be in error, radiometric dating validity is further challenged.

The biases of biblical theology in regard to creation and scientific assumptions and experimentation in regard to creation are usually viewed as being totally at odds with each other. And one bias viewed as holding more credibility than the other. Thus, in real terms; one is viewed as 'right' and the other is viewed as 'wrong'. Historically, prior to the around the 1700s the bias of biblical theology was viewed with consensus as the more credible viewpoint regarding creation. Subsequent to that time frame the consensus viewpoint has shifted viewing that science holds the most credible viewpoint regarding creation. Yet within their biases, there are details of creation overlooked or rejected because the details don't conform to their biases. The three most contentious and specific details which are non-conforming are: man, the earth was the first created body of creation of the universe, the element of the passage of time.

Orthodox biblical theological bias states creation of a single man-woman unit commonly referred to as Adam & Eve. While scientific evidence points to the multiplicity of man. The disparity between these biases was brought forth within the writings contained herein in the Chapter M.an and validated the multiplicity of man. Man (Adam) was created on the third day. Eve was created on the sixth day; and men and women extraneous to Adam & Eve were created on the sixth day.

The creation of the earth as the first created body of creation within the bias of biblical theology is supportable within the bias of science. While earth is the platform for scientific observation of the cosmos, if the Big Bang theory is valid; it is probable that the earth is the point of singularity. According to biblical bias the earth was created first then the sun. moon and stars were created subsequent to the earth [earth on the third day of creation and sun, moon and stars on the fourth day of creation].[168] Science has recognized via redshift observations and calculations that almost all galaxies in all directions are moving away from earth. Astronomer Edwin Hubble attributed these observations to the expansion of the entire universe. Yet, from a practical point, if the point of singularity were somewhere else in the universe other than earth or the vicinity thereof there should be observations of significant numbers of blueshifted[169] galaxies moving toward earth. But, even most of the blueshift galaxies are only moving toward earth in a relative manner as their blueshift indicating motion towards earth is a sum of their doppler blueshift and their cosmological redshift wherein the blueshift wins.[170] In reality the exact location of the earth in the universe is unknown. Based on observations of other galaxies scientific bias assumes that earth is located remotely in an arm of a spiral galaxy named the Milky Way. Theological bias historical assumed that earth is the center of the universe.

The passing of time since creation between biblical theological bias and scientific bias is seemingly a difficult bias to overcome. The biblical bias based on recording events of history from the first day of creation to the present provides a time from creation to the present of around 6,000 years. The

---

168 Genesis 1:9-19

169 Blueshift is spectral analysis of elements showing a shift to the blue spectrum of light and indicates movement towards the observer.

170 https://www.astronomy.com/science/do-blueshifted-galaxies-contradict-the-expansion-of-the-universe/

scientific bias based on theories and refined theories as the result of observations and experimentation indicates an elapsed time since creation of about 13.8 billion years. This bias from either perspective might thought to be difficult to reconcile. Even with the example shown previously of diamonds containing carbon-14, the age of diamonds was calculated to be about 55,000 years old. Although the radiometric dating of the diamond samples showed a reduction in age from billions of years to 55,000 years: the refined age still significantly surpasses the biblical bias of roughly 6,000 years.

The discrepancy in time since creation between the biblical theological bias and the scientific bias might have some degree of reconciliation in the example of a disease commonly referred to as 'old man disease'. The medical name for 'old man disease' is Progeria (Hutchinson-Gilford progeria syndrome). According to information from the Mayo Clinic, Progeria is an extremely rare genetic disorder that causes children to age rapidly starting in the first two years of their life. They do undergo physical growth. During the aging process the child's motor development nor their intellectual development is affected. But, this rapid aging is observable in all other aspects of their being. They get wrinkles, their hair falls out, they develop dental problems, they lose their hearing, they get arthritis. And, generally they develop all the ailments associated with the aged. Furthermore, the result of these symptoms is that their appearance is observably that of being an old person. This rapid aging is not just the appearance of aging but each cell in the body undergoes the natural aging process at an accelerated rate. Current medical literature shows the cause of Progeria to be a single gene. Mayo Clinic reports, "A change in one gene causes progeria. This gene, known as lamin A (LMNA), makes a protein that's needed to hold the center of a cell, called the nucleus, together. When the LMNA gene has a change, a flawed lamin A protein called progerin is made. Progerin makes cells unstable and appears to lead to progeria's aging process."[171] There is no certainty when the lamin A gene undergoes change of becomes flawed. The maximum age children with Progeria may reach is about 20 years old. A 20 year old child may look like a 90 year old adult, Or, a 10 year old child may look like a 90 year old adult. A child with Progeria has a known date of birth but, as they grow they age rapidly. Accordingly, the age of life does not reconcile with the observable age.

Such an example explains the time elapsed since creation as viewed from a biblical theological bias and a scientific bias. From biblical records and accounting, the biblical theological bias, creation has a specific age – roughly 6,000 years. From a scientific bias creation has an observable age – roughly 13.8 billion years. Just like a child with Progeria the actual age since birth does not match the observable age of the child: the actual age of creation does not match the observable age of creation. Progeria is caused by a single gene. Likewise, there is a recognizable single event in creation that lends itself to be the cause for the actual age of the universe to not match the observable age of the universe.

From the biblical narrative of creation in the book of Genesis creation of the physical elements of the universe are all declared "good" upon their creation. Prior to God resting (from his creative work of creation of this universe) on the seventh day[172] all was good. However, at the event of Adam's disobedience to God when he ate of the fruit of the tree of the knowledge of good and evil: things changed! It is recorded that: the serpent was changed and locomotion would be upon its

---

171 https://www.mayclinic.org/diseases-conditions/progeria/symptoms-causes/syc-20356038#:-:text=Progeria%20(pro-JEER-e-uh).
172 Genesis 2:2

belly; the woman would experience sorrow in child birth; there was enmity created between the seed of the woman and the seed of the serpent; man would cultivate among thorns and thistles and the ground was cursed because of man; God slayed animals to make coverings for Adam and his wife;[173] and, Adam and his wife were sent forth out of the garden of Eden. This much we know about the immediate ramifications of Adam's disobedience, or sin. It is not directly recorded as to other ramifications which resulted from Adam's sin. Pastor Chris Basham of The Church at Odessa in Odessa Florida emphasizes in sermons about the unknown and obstensibly far reaching ramifications of personal sin. Yet, the sin of Adam had far reaching ramifications beyond its immediate consequences. Further than perceived within theological bias. Sin caused corruption in creation. That corruption affected not only man but the entirety of creation. Upon referencing back to the concept of the 'butterfly effect'; anything affects everything. In the book of Romans, chapter eight, verse 22 it is written, *"For we know that the whole creation groaneth and travaileth in pain until now."* Contextually this is in relation to the reconciliation of man to God through the redemption of Jesus Christ. It is due to Adam's sin that this reconciliation is necessary. But, this groaning and travailing indicates things are not right or in order in the entirety of creation. This single event of Adam's sin, like the single gene causing Progeria created a division between the known and the observable. Like a child with Progeria with a known date of birth: the known time since creation is just over 6,000 years. Like a child with Progeria and the observable appearance of an old man and each and every cell observable of age: the observable time since creation is 13.8 billion years. Within this theory, the biblical theological bias as to the time since creation is reconciled with the scientific bias as to the time since creation.

---

173 Genesis 3:21

The Bible is referenced as not being a "science book" and the scientific method is viewed as not being "theology". However, each of their biases overlap to provide a complete picture. In reality, this viewpoint does unconsciously exist. The two biases – biblical bias and scientific bias can be seen as two separate observers looking at a simple classical optical illusion, the Rubin Vase. However, the Rubin Vase is not truly an optical illusion but, is the difference in an observer's perception from that of another observer. The Rubin Vase is viewed by one observer as being just a vase with a nondescript or even an elaborate background. However, another observer sees not a vase but, the faces of two people facing each other and what the the other observer viewed as a vase is seen as the background. A simplistic rendering of a Rubin Vase:

The vase/faces and the background are all a part of the entire picture. This overlap of the biblical bias and the scientific bias can be seen in some selected quotes from renowned scientists:[174]

"The really amazing thing is not that life on Earth is balanced on a knife-edge, but that the entire universe is balanced on a knife-edge, and would be total chaos if any of the natural 'constants' were off even slightly. You see," Davies adds, "even if you dismiss man as a chance

174 https://www.simpletoremember.com/articles/a/science-quotes/

happening, the fact remains that the universe seems unreasonably suited to the existence of life—almost contrived—you might say a 'put-up job'." - <u>Dr. Paul Davies (Professor of Theoretical Physics at Adelaide University)</u>

"If you equate the probability of the birth of a bacteria cell to chance assembly of its atoms, eternity will not suffice to produce one... Faced with the enormous sum of lucky draws behind the success of the evolutionary game, one may legitimately wonder to what extent this success is actually written into the fabric of the universe." - <u>Christian de Duve. (Nobel laureate and organic chemist)</u>

"As a man who has devoted his whole life to the most clear headed science, to the study of matter, I can tell you as a result of my research about atoms this much: There is no matter as such. All matter originates and exists only by virtue of a force which brings the particle of an atom to vibration and holds this most minute solar system of the atom together. We must assume behind this force the existence of a conscious and intelligent mind. This mind is the matrix of all matter." - <u>Max Planck (founder of the quantum theory and one of the most important physicists of the twentieth century)</u>

"The statistical probability that organic structures and the most precisely harmonized reactions that typify living organisms would be generated by accident, is zero." - <u>Ilya Prigogine(Chemist-Physicist) Recipient of two Nobel Prizes in chemistry</u>

"We are, by astronomical standards, a pampered, cosseted, cherished group of creatures.. .. If the Universe had not been made with the most exacting precision we could never have come into existence. It is my view that these circumstances indicate the universe was created for man to live in."- John O'Keefe (astronomer at NASA)

The Bible provides a scientific basis in its text from which scientific observation and experimentation should flow. Albert Einstein is quoted, "The human mind is not capable of grasping the Universe. We are like a little child entering a huge library. The walls are covered to the ceilings with books in many different tongues. The child knows that someone must have written these books. It does not know who or how. It does not understand the languages in which they are written. But the child notes a definite plan in the arrangement of the books - a mysterious order which it does not comprehend, but only dimly suspects."[175] In the television series the X-files which ran from 1993 to 2002 two FBI (Federal Bureau of Investigation) special agents attempted to solve crimes involving UFOs (Unidentified Flying Objects), the paranormal and government conspiracies. In one episode the female agent Skully said to the male agent Mulder, "The truth is out there but, so are lies."

Because man has difficulty grasping God and his greatness and power, billions of years since creation resonates in the human mind; yet God shows these attributes of his greatness and power in creation. Directing attention to the text of the Bible and establishing it as truth regarding creation is a basis from which the child in Einstein's quote would be grounded; and begin to understand the contents of the multitude of books in the vast library. And, directing attention to the text of the Bible and establishing it as truth regarding creation would allow avoidance of the lies referenced in the quote from the X-files. Such realignment of perceptions would dissolve the biases between

---

175 https://www.simpletoremember.com/articles/a/science-quotes/

the religious community and the scientific community. More importantly though, a new era of discovery and enlightenment would happen, To God be the glory.

www.ingramcontent.com/pod-product-compliance
Lightning Source LLC
Chambersburg PA
CBHW021112090426
42738CB00006B/608